from FIRE *to* RUST

Business, Technology and Work at the Lackawanna Steel Plant, 1899-1983

Thomas E. Leary
Elizabeth C. Sholes

Buffalo and Erie County Historical Society

This book is dedicated to the men and women of the Lackawanna steel plant

Publication of this book was supported by a grant from the New York State Council on the Arts.

First Edition

Library of Congress Cataloging in Publication Data

Leary, Thomas E. 1946-

From fire to rust.

Bibliography: p.

1. Lackawanna Steel Company — History. 2. Bethlehem Steel Corporation — History. 3. Steel industry and trade — United States — History. I. Sholes, Elizabeth C., 1946- II. Title. HD9519.L2L43 1987 338.7'669142'0974796 87-692
ISBN 0-939032-00-7

INTRODUCTION

During the summer of 1919 a young man named Charles R. Walker went to work in a steel mill. Walker's stay at the plant was temporary; he subsequently became a scholar famed for his sociological studies of factory life. But during that season after a great war and before a great steel strike, he cast his lot with half a million other men by entering the complex, dangerous, and fascinating world of a 20th-century steel plant.

Looking backward and forward from Walker's vantage point, the history of business, technology, and work in modern steel production appears as a series of both cataclysmic and incremental changes. This book attempts to increase public understanding about the history of steelmaking by taking readers inside a major 20th-century plant in Lackawanna, New York. Discussions of corporate policies, production technology, and industrial authority shed light on the particular experiences of the men and women who labored there but also represent many developments typical of the steel industry as a whole.

Charles Walker and his contemporaries performed jobs that had undergone some radical transformations during the decades surrounding the turn of the century. The most visible symbols of the new workplace were the enormous machines that had replaced men at many of the production stages necessary for wresting steel from lumps of minerals.

> The mechanizing of processes in the past two decades has revolutionized the industry, each new machine displacing skilled men at the top and unskilled at the bottom. More and more the steel job tends to become the job of a machine, each new machine tending to abolish either the occupations of a dozen common laborers in favor of one semiskilled man or of a few skilled men in favor of one not so skilled.

Inside the mills during young Walker's time, another change was brewing. Corporate and plant management had consolidated a rigid system of arbitrary authority over critical decisions affecting steelworkers' jobs and lives. Some steelworkers in the 1870s and 1880s had exercised partial control over aspects of their work because of their skills or the influence of a craft-based union, the Amalgamated Association of Iron, Steel and Tin Workers. However, new production processes and ethnic divisions undermined key sources of 19th-century workers' strength; the most dramatic example of the union's decline was its defeat in the 1892 lockout at Andrew Carnegie's steel works in Homestead, Pennsylvania. The 20th-century labor movement in steel had to be rebuilt on new foundations: the job structure associated with contemporary technology and the political maturation of immigrant workers. By 1919 the men who worked in the steel mills were prepared to seek new means of counterbalancing their employers' monopoly of power over the workplace.

Big, integrated mills that made iron and then refined it as steel contained a multitude of distinct occupational environments. As veteran steelworker Charles Spencer has written:

> A steel mill has a broad mix of occupations and workers, skilled, semiskilled and unskilled; Black, white, Latinos; ethnics and WASPs; precision workers and assembly-line workers; backbreaking labor and light machine work; outside and inside work; transporting, shipping, fabrication, forging, smelting, constructing; simple mechanical and sophisticated computer processes; it's all there in an integrated steel plant . . .

Another dimension to the story is the history of the corporation. The work which Walker and Spencer encountered was shaped in whole or in part by the dictates of the large business enterprise under the control of its directors. The contemporary corporation was also recast from its 19th-century counterpart. The latter-day steel firm is no longer an entrepreneurial venture such as it was under Carnegie; rather, it is a modern business corporation employing massive infusions of outside capital and creating its policies within the larger network of interlocking relations with other businesses.

The corporation draws its strength from the law. Under interpretations of the 14th Amendment of the Constitution, courts have provided the company with the status of a fictitious person and have granted it all the rights of a human being. Stockholders are thus protected from liability for corporate actions and freed from the day-to-day responsibility for its operation. On the other hand, while corporate officers are supposedly trustees for the stockholders' interests, the "business-judgement rule" allows management to run its own show. As long as decisions are made "in good faith" managers are immune from stockholder retaliation if things go awry.

Corporate directors are thus the sole arbiters of the company's interests. They may be employees or even owners, but in either case they are likely to have one of two interests in the company. They might be the men (rarely women) who founded the business in order to have it function as a going concern, or they might be directors whose interests in the company are restricted to the value of the company's stock and its dividends. These two positions are often contradictory, but the legal right to direct the fate of the company is vested in all directors equally, regardless of philosophy.

Control over the business extends beyond the executives who work for it, the "inside directors", to include a welter of interested parties — banks, customers, suppliers, institutional stockowners, even competitors — any one of whom is potentially eligible to sit on the Board as an "outside director". Outside directors help shape company policy beneficial to their mutual interests. In 1914 the Clayton Act, an outgrowth of the Progressive Era anti-trust movement, forbade directors of competing companies from serving on each other's boards. However, nothing prevents competitors such as Bethlehem Steel and U.S. Steel from sending their directors to sit together on the board of a "neutral" third-party corporation.

Often centered around powerful banks, companies have used these indirect ties to build important affiliations or "cliques" through which they could create consensus policies. They could and did carve out non-competitive market and pricing agreements in contravention of supply-and-demand and, occasionally, the law. By preserving these alliances to promote friendly relations with key banks, captive suppliers, and dependent customers, America's largest firms weathered the Depression, two world wars, and recessions. If not precisely monopolistic, these cliques contributed to oligopoly.

Ironically, the very strength of decades-old alliances paralyzed industries' capacity to be flexible in finding solutions to post-1970 economic problems. The tight interdependence among banks, manufacturers, and institutional investors made it virtually impossible to lower prices, husband resources, and generally weather crises with conservative, long-range plans. Lenders, bond holders, and stock owners all demanded persistent payments to bolster their own battered economic positions. By 1980 corporations such as Bethlehem believed they had only two options to increase profits: a radical assault on wages and work practices and the liquidation of operations.

In sum, the changes we will see in technology and authority at the Lackawanna steel plant reflect over 83 years of policies and priorities which were characteristic of 20th-century capitalist practice as a whole.

Chapters 2, 5, and 8 examine changes in iron and steel technology in some detail for the benefit of former Lackawanna employees familiar with the plant as well as readers interested in comparing its operations with other plants. For those unfamiliar with the principles of the basic processes for making iron and steel, a brief introduction to terminology used in the text may be helpful. The processes involve the application of heat or pressure to alter the chemical composition and physical structure of materials. The discussions that follow condense the many steps involved by focusing on four key operations: coke ovens, blast furnaces, steelmaking, and rolling mills. We will inventory these areas over different time periods to compare changes in production techniques. Pictures accompanying the text are generally arranged to follow this elementary production sequence.

Coke is used in the blast furnace as a fuel and as an agent in making iron from ore. The production of coke takes place in by-product ovens where bituminous coal is baked in narrow, airtight chambers. Some of the gases and vapors expelled from the coal at high temperatures are recycled to heat the ovens and fuel other operations; the remainder is reprocessed to obtain chemicals.

Blast furnaces produce pig iron, the intermediate form of metal from which steel is derived. On the skyline of a steel plant the silhouettes of these towering structures stand out as the most impressive architecture. Iron ore, coke, and limestone are dropped into bells at the top of the furnace and descend through an inferno of heat and gases to emerge at the bottom transformed into molten iron, also called hot metal. The chemical reactions inside the furnace also produce liquid slag, the vehicle that carries impurities away from the iron. Blast furnace pig iron lacks the tensile strength, elasticity, and malleability characteristic of steel and therefore requires further refinement.

Steelmaking includes the purification of pig iron in conjunction with melting other ingredients such as pieces of scrap; materials called fluxes that form the cleansing slag are also blended into the recipe. Processes used for cooking up large tonnages of steel include the Bessemer converter, the open hearth furnace, and the basic oxygen furnace (BOF). Molten steel produced by any of these methods is poured, or teemed, from a ladle into molds where it solidifies as an ingot.

Rolling mills, like houses, feature many different individual details while sharing certain essential characteristics. All rolling mills use pressure to shape steel into semi-finished and finished forms. A rolling mill consists of individual roll stands, usually connected by tables that carry the steel to and from the rolls as well as between stands. Each stand contains 2, 3, or 4 rolls mounted one above the other in a single housing; depending on the roll arrangement, mills are classified as 2-high, 3-high, or 4-high. The rolls themselves are grooved or smooth, depending on the mill's product line. Paired upper and lower rolls revolve at equal speeds but in opposite directions so as to draw the steel through. The size of most mills, such as a 44" blooming mill, is expressed in the number of inches between the centers of the pinion gears that transmit power to the rolls.

Steel is usually rolled when hot so it will be more plastic. Prior to rolling ingots, their temperature is regulated in soaking pits where the steel is bathed in heat. The ingot is then rolled into one of several semi-finished forms on the primary mills that produce blooms, billets, or slabs. Semi-finished forms vary in length and cross-sectional area. Surface defects may be treated through assorted conditioning methods such as scarfing, chipping, or grinding. The temperature of semi-finished steel is adjusted again in reheating furnaces before being rolled into rails, structurals such as beams, bars, plates, strip and sheet, or other finished products.

Processes conducted on this scale require prodigious expenditures of heat and energy. Measures to recover waste products that might otherwise be lost constitute an important part of steelplant engineering. Two common types of heat conservation, the regenerative and recuperative methods, both reuse waste gases. Each involves preheating the air or fuel required in maintaining combustion at coke ovens, blast furnace stoves, open hearths, soaking pits, and reheating furnaces. In practice the recycling of heat and energy links together the various departments of a steel plant as closely as the mechanical systems that transfer raw materials or products.

Over the years the introduction of new equipment and the obsolescence or persistence of older methods affected not only the plant's productive capacity but also its structure of jobs, skills, and work processes as well. It is this fluid combination of change and continuity that we hope the capture in the ingot molds of prose.

By the time Charles Walker entered the mills, the geographic center of the steel industry had shifted along with the business context, technological base, and division of labor. Turn-of-the-century corporate officials calculated that the prime locations for future steel plants would be found along the Great Lakes rather than in the congested river valleys of Pennsylvania and Ohio were the mills were born. The westward drift of steel markets toward the growing industrial centers of the Midwest combined with changes in the sources of fuel and ore to enhance the attractiveness of lakeside locations. Reductions in the fuel consumption rates of blast furnaces, plus the introduction of alternative energy supplies from by-product coke ovens, meant that the new integrated steel plants could loosen the ties that had bound their predecessors to the inland coal regions. Also, sites where lake vessels could unload iron ore became more attractive. Iron had begun moving upon the waters in the mid-19th century. By the 1890s the rising tide of ore deliveries was accelerated by new deposits discovered around the upper Great Lakes, especially at the Mesabi Range in Minnesota.

One company that planned to capitalize on the new economies of lakeside location originated in Scranton, Pennsylvania. Since the 1840s when nearby anthracite coal had been the best fuel available for blast furnaces, Scranton had ranked as a major ironmaking district. The Lackawanna Coal and Iron Company of Scranton joined the first wave of Bessemer-steel rail producers in 1875, and the Scranton Steel Company erected a second Bessemer plant that began rolling rails in 1883. These two operations merged to become the Lackawanna Iron & Steel Company in 1891.

By the close of the 1890s, the westward movement of customers and raw material supplies had drained away some of Scranton's locational advantages. Short-term competitive adversity and projected strategic limitations provoked a drastic response from Lackawanna Iron & Steel. In 1899 it announced that steel operations at Scranton would cease and a new plant would be built near Buffalo, the lake port the company had used for several years when shipping rails to Duluth.

The consequences of the shutdown were devastating for Scranton, though many skilled steelworkers were relocated to New York along with the plant. When the north and south mills were closed in Scranton during 1901-1902, 2500 jobs evaporated. A newspaper article comparing the construction at Buffalo with the destruction at Scranton observed:

> The work in progress gave a somewhat sad picture of the decadence of Scranton . . . The plant is now being removed here. The machinery in the mills has been shipped here, buildings have been torn down carefully and the material has been sent here. There are great piles of window frames, door frames, roof beams, etc. at 'Stony Point' now. The mills torn down at Scranton are rising again, almost like magic, at South Buffalo.

Through the decades that followed Charles Walker's stint in a steel mill, business, technology, and work continued to change. Amid these developments the Lackawanna plant under Bethlehem Steel Corporation remained one of the nation's premier installations through the 1960s. The people employed there experienced firsthand the events that students of history should recall in years to come. But the course of history eventually sapped the vitality of Lackawanna, and the decisions of men terminated steelmaking after 80 years. The fires in the heart of the plant died out, leaving only corroding rust as a monument.

ACKNOWLEDGEMENTS

A great many people contributed to this project. Without their help we would never have been able to bring this book to completion. Any errors are, of course, ours alone. We are indebted to the present and former Directors of the Buffalo and Erie County Historical Society, Dr. G. Rollie Adams, Dr. Paul Bulger, and Dr. William Siener. Former and current library staff provided cheerful and tireless assistance, and we thank Dr. Herman Sass, now retired, Mary Bell, Molly Howard, and Yvonne Foote. Interpretive, esthetic, and moral support were also provided by Education Coordinator Mary Cronin and Photographer Thomas Payne.

The staff of Bethlehem Steel District Reclamation were the backbone of this project, providing invaluable historical and technical help and access to the plant and records. We give special thanks to Eion Gordon, Paul Pietrzak, and Pat Schamann and to Coke Oven Supervisor Joe Murphy as well as to plant clerical and security staff who made our research easier and kept a protective eye on us.

We owe an enormous debt to the many current and past employees who taught us how things worked at various mills and lent their time and knowledge to improving our own. Our deepest thanks to Mr. and Mrs. Don Landsittel, John Quider, Sam Ruggiero, Dick Mitchell, Ed Sobala, Mrs. Kozub, Esther Kowal, Art Sambuchi, Pat Danieu, Jack Williams and others who wish to stay anonymous or whose names, regrettably, we never learned. Before we ever began this book, the groundwork for preserving steelworkers' history was laid by two dedicated people, Jerry Soltis and Jim Swinnich, who founded the Steel Plant Museum at the Lackawanna Public Library. Mia Boynton extended the record with the first steelworkers' exhibit, *In the Shadow of the Mills*, which toured Erie County during 1984-1985 for BECHS.

We would also like to thank the staff of the Butler Library at the State University College at Buffalo, particularly our dear friend Sr. Martin Joseph, Archivist and trustee of the BECHS *Courier-Express* collection, plus Mary Delmont, and the reference and microfilm staffs. Thanks also to the superb reference staff at Buffalo and Erie County Public Library whose collections are extraordinary and whose help was invaluable.

Much of the research for this book was conducted as part of a cooperative project with the Historic American Engineering Record, National Park Service, Department of the Interior. HAER sent photographer Jet Lowe here in 1985 and 1986 to document the Lackawanna steel plant for the Library of Congress which he did with extraordinary skill. We also thank Dr. Robert Kapsch, Bill Lebovitch, and Richard Anderson of HAER for believing in the importance of Buffalo's industrial history. We are indebted to Susan Turner who so ably coordinated much of the HAER recording project and to Bonnie Ott, Instructor at SUNYAB School of Architecture and her three graduate students, Francisco Hidalgo-Romero, Carolyn Honan, and Rob Hayes who delineated the HAER maps and drawings used in this book.

Photos were also furnished by the New York State Museum Archives, thanks to Duncan Hay, and by the Hamburg (NY) Historical Society thanks to Town Historian Joe Streamer. General background was provided by the Malloy family, owners of the Harbor Inn and local historians without peer. Thanks to Paragon Business Service who typed much of this manuscript and did it well.

We owe our deepest personal thanks to our dear and loyal friends, Addie and Murray Levine, and to Bob O'Connor and Mary Jo Laughlin. Above all we extend our love and thanks to our families, DeVer and Louise Sholes, Mildred Leary and Glenn Cutcliffe.

Thomas E. Leary
Elizabeth C. Sholes
Buffalo, NY

LACKAWANNA STEEL COMPANY
LACKAWANNA STEEL COMPANY
LACKAWANNA STEEL COMPANY
LACKAWANNA
STEEL SHEET PILING.

LACKAWANNA STEEL 1899-1922

1. THE CORPORATION

In the Spring of 1899, a group of prominent Western New York businessmen met at the exclusive Buffalo Club on Delaware Avenue. The purpose of the meeting was to complete plans for enticing Lackawanna Iron & Steel Company to the shores of Lake Erie by meeting the Scranton, Pennsylvania company's request for capital, land, and dockage facilities. Local capital raised in that historic meeting was substantial — $2.5 million out of $15 million total — involving subscriptions from several prominent area capitalists led by John J. Albright, John G. Milburn, and Edmund Hayes. Albright had already acquired rights to 1500 acres south of the city under the nominal ownership of the newly-incorporated Stony Point Land Company; by December, 1899 he would buy still more from the city of Buffalo, then resell it to the land company at a profit of $644,154 to himself. Stony Point Land was a designated agent for Lackawanna Iron & Steel.

North Office Building, Lackawanna, NY. Built in 1901, this edifice was likened to the Buffalo Historical Society for architectural grandeur.

The establishment of Lackawanna Steel, its New York name, "meant as much for Buffalo as the Carnegie interests did for Pittsburgh," declared Milburn. Over more than eight decades, the burgeoning iron and steel plant did, indeed, live up to Milburn's prophesy. But the company began its sojourn in Western New York much as it finally died, as a runaway industry abandoning jobs and its founding community when profits lagged and costs increased. In between, the fortunes of Lackawanna Steel and its successor, Bethlehem, were both tempestuous and dazzling.

Business History

Once the company made its transition from Scranton to Buffalo in 1902, not only did the plant site develop rapidly but so did the surrounding area. Coming hard on the heels of the city's 1901 Pan-American Exposition, the swift growth and sheer size of the new mill complex inevitably invited comparison. As one local banker told a Buffalo newspaper:

> Excellent as is the Pan-American enterprise for the development of our local industries, the establishment of a mammoth steel plant which with its allied industries will give much needed employment to thousands of men, is likely to be a much more substantial factor in the city's growth and permanent benefit.
>
> The simultaneous coming of the Pan-American and the steel plant does not by any means dim the glory of the latter in the eyes of the commercial world.

Even the construction of the company's ornate North Office Building drew favorable praise:

> It looks more like a museum, even suggesting in general outline the building of the Buffalo Historical Society in Delaware Park.

The greatest interest in Lackawanna Steel, however, lay in its productive promise and its proffered hopes for jobs and prosperity.

From the time of its formal incorporation, 14 February 1902, Lackawanna Steel set its expansion plans along all four lines of corporate growth: horizontal — building

Lackawanna Steel Co. trade show display (1909).

one primary product line; backwards vertical — expanding control over raw materials such as fuel and ores; forward vertical — developing sales and shipping facilities; and diversified — creating product lines outside the main corporate manufacture. Lackawanna's primary focus was, of course, on horizontal growth. The Lackawanna Iron & Steel company already had sizeable blast furnace operations in Lebanon, Pennsylvania plus the Franklin Furnaces in New Jersey and two Bird-Coleman furnaces in Cornwall, Pennsylvania. The company's primary reason for settling in the Buffalo area was to expand its basic capacity to "manufacture iron, steel, and other metals" and "to erect and operate furnaces, forges, mills, foundries, and other manufacturing establishments" necessary to primary metal production. The combined productivity figures for the New York site alone were impressive. By 1920 the company's 9 blast furnaces had an annual capacity of 1.4 million tons of pig iron, while the 4 Bessemer converters and 24 open hearths yielded 1.8 million gross tons of ingots. Finishing mills, including the rail mill, annually produced 1.5 million tons while the 469 coke ovens generated 700,000 tons of coke plus by-products. By 1922 when Lackawanna Steel was purchased by Bethlehem, the smaller company was the nation's fourth largest producer, manufacturing 3.65% of all U.S. capacity. Compared with the industry giant, U.S. Steel, Lackawanna's capitalization was only 1/35, but its output was 1/8, making its earnings on investment obviously much higher.

Lackawanna Steel, along with its parent, Lackawanna Iron & Steel, owned much more than just its Pennsylvania and New York sites. The company recognized early on the importance of capturing sources of fuel and ore to minimize costs of purchasing these items from outside suppliers, and therefore engaged in substantial development of backward-vertical integration. Through a 99.9%-owned subsidiary, Lackawanna Coal & Coke, the steel company owned coal lands in Pennsylvania. With similar majority holdings they also controlled the Tilly Foster Iron Mines in New York, the aforementioned Franklin Iron Company and its mines in New Jersey, and the Scranton Mining Company in Minnesota's vast Mesabi iron range.

These controlling shares in raw materials subsidiaries were supplemented by lesser holdings in other raw goods operations. Lackawanna controlled 33.3% of the Witherbee, Sherman & Co. mines in Port Henry, New York and a similar interest in the Lebanon Cornwall Ore Bank in Pennsylvania. They also owned 27.5% of Odanah Iron Company in Wisconsin's Gogebic Range, 18.7% of Verona Mining in Michigan (which increased to 50% ownership by 1920), 66.6% of Hobart Iron Co. and 61.1% of Corsica Iron Company, both in Minnesota, and a 59.2% share of West Virginia's Ontario Mining Company which held leases in the Mesabi Range. According to the industry journal, *Iron Age*, as early as 1903 these "shares of ore and mining companies . . . together with certain ore contracts" owned exclusively by Lackawanna Steel "represent the ownership or effective control of upwards of 56,000,000 tons of iron ore." In 1907 the company's control over fuel reserves was supplemented considerably by the incorporation of their Ellsworth Collieries Company which held 15,000 acres of coking coal lands with 84 million tons of reserves and an annual 4-mine capacity of 2 million tons. Ellsworth provided enough limestone, critical to blast furnace and open hearth production, to sustain the firm for 30 years.

Recognizing the potential costs of deploying the firm's major product lines throughout the nation, Lackawanna quickly developed several forward-vertical holdings designed to reduce those expenses. Among their holdings were the South Buffalo Railway which transported materials into, around, and from the mills, tying the steel plant in Buffalo to other rail lines which could ship goods to the markets of the east and midwest. In addition they formed the Seneca Transportation Company for water shipments. Over their 23-year life span the company continued to develop domestic and export sales firms as well as a more general retail merchandizing operation.

Lackawanna Steel diversified in small but important ways. They owned water companies, a real estate development firm, and even a bank in Ellsworth, Pennsylvania. Finally, as with most companies, they made portfolio investments in other corporations which were neither subsidiaries nor under their control. As stockholders they earned small but steady income from their investments: $229,374 in 1901 and $451,429 in 1910.

Ownership and Control

In the early days after the company's 1899 organization at the Buffalo Club, Western New York investors played a substantial role in Lackawanna Steel's affairs. Local investors held a significant block of company stock, owning 16.7% or $5 million out of $30 million worth of shares. The Board of Directors, the company's policy-planning body, had 2 out of 23 directors from Buffalo, John J. Albright and Edmund Hayes. However, after 1910 when the directors expanded the number of shares that sold below par to generate new capital, Buffalo investors — among the first in the company — were left with overpriced, low-income stocks. Consequently, local investment fell sharply as faith in the company declined.

Simultaneously, the composition of the Board, always impressive, began increasingly to represent companies and capitalists operating at the highest levels of corporate America. By 1920 only one Western New Yorker, Albright, remained on the Board as directors of the caliber of Cornelius Vanderbilt III and prominent bankers became more prevalent.

The primary tie which all corporations sought to cultivate was to banks, with railroads running a close second. Lackawanna was no exception, and men drawn from these two important business sectors dominated the Board in its early as well as later days. In 1903 Lackawanna's first Board of Directors included men representing 33 separate banks and 36 different railroads. In 1920, the last prosperous year before the merger with Bethlehem, the number of these critical ties had declined to 22 banks and 27 railways, but the sheer quality of affiliations had not diminished. Throughout its life-span, Lackawanna was closely allied to the two most important business circles of the time: the Morgan group, headed by banker J.P. Morgan and his son, and the Rockefeller-Stillman-Kuhn, Loeb-Harriman clique.

In both years, 1903 and 1920, National City Bank was represented on the Lackawanna Board. The bank itself united both cliques as it was owned by Rockefeller associate James Stillman but was itself tied through common directors to the Morgan banking operations as well as the Kuhn, Loeb private investment bank. Lackawanna's Chairman of the Board, Moses Taylor, was a director of National City Bank since his grandfather was the bank's founder; other National City directors tied to Lackawanna were Henry A.C. Taylor and Stephen S. Palmer.

In addition to National City, several other crucial banking ties were present on the Board. Among these were U.S. Trust, National Bank of Commerce, Chase National (the precursor to Chase Manhattan, the central Rockefeller bank), and Guaranty Trust. The latter, while not yet merged with J.P. Morgan, was a close Morgan ally. Private investment bankers were equally well-represented in the persons of, first, D.C. Blair and then his son, Clinton, plus Adrian Iselin and Darius O. Mills.

The most important banking ties Lackawanna maintained were those to the banks which had provided the steel company with its basic capital in the form of loans. Farmers Loan and Trust had provided the company with $21 million and consequently had a director present to oversee steel operations. While not a major New York bank at the time, it later would merge with the much larger National City Bank, the forerunner to today's Citicorp. Two other financial trustees for Lackawanna debts had seats on the Board of Directors, Central Union Trust and a Philadelphia bank, Commercial Trust. By 1920, the banking interests helping to create policy for Lackawanna were truly national in scope, including Brown Bros., the private investment bank which merged with the powerful Harriman interests; New York Trust; Hanover National, the original incarnation of the contemporary Manufacturers Hanover Trust, and even the California financial power, Wells Fargo.

The relationship between steel companies and railroads was an old, established one. Steel companies relied on close alliances with railroads to obtain favorable rates and shipping routes which would minimize transport costs of both raw goods and finished products. The rail lines, in turn, wanted as much business as possible from the manufacturing sector and courted steel companies vigorously. Therefore, the presence of large numbers of directors from railroad concerns on the Lackawanna Board should be of no surprise.

The key railroads were those tied to the Morgan or allied cliques. The Erie Railroad plus three of its subsidiaries, and the Great Northern, Union Pacific, and the Buffalo, Rochester & Pittsburgh lines all had ties to Lackawanna Steel and were central railway holdings of major capitalist combines. In addition, the Delaware, Lackawanna & Western was one of the family operations of Lackawanna director, Cornelius Vanderbilt III. Other directors were from the New York Central and its two subsidiaries; the Atchison, Topeka & Santa Fe; the old "Nickel Plate" line, the New York, Chicago & St. Louis; the Missouri Pacific; the Chicago & Alton, and the Louisville & Nashville. Not only were the railroads important carriers they were also the owners of strategic subsidiary properties, controlling reserves of coal and ore which were possible sources of cheap raw materials for Lackawanna.

In 1903, before the 1914 Clayton Act forbade interlocking directorships between two theoretically competing firms, Lackawanna's directors included other steel and iron firm representatives. Clearly these relations were potentially collusive, allowing ample opportunity for price-fixing and non-competition, but they were so valuable that even after the federal government's prohibition on such relations, Lackawanna managed to preserve some key alliances. In 1903 Earl Dodge was on the steel company's Board. Not only the head of Phelps-Dodge mining interests, Dodge was also a director of U.S. Steel. In that same year Lackawanna interlocked with another steel finisher, Shenandoah Steel Wire located also in Lackawanna, New York, and with Harvey Steel Company. Lackawanna drew board members from mining concerns theoretically in competition with Lackawanna's subsidiaries. By 1920 such relations had unquestionably been reduced though by no means eliminated. Directors from Pressed Steel Car, Adrian Furnace, Burden Iron Company and similar enterprises continued to be present on the Buffalo company's Board.

As the company prospered over its good years, the Board recruited still more important business representatives. The 1920 directors tied Lackawanna to potential customers such as Pullman and International Harvester and to large conglomerates and utilities such as AT&T, American Express, and Commonwealth Edison of Chicago. These ties are important for two reasons. First, they measure Lackawanna's stature within the larger corporate community, and second, they illustrate the wide range of influences setting corporate policies and thus establishing the direction of plant operations.

Beneath the Board of Directors was a management structure which oversaw actual steel production. The top executive management remained at company headquarters, in New York City, leaving day-to-day operations to on-site plant managers. However, these steel men (and mining experts in the case of ore and coal facilities), had no say over the rate and direction of capital expansion established by the Board or even over marketing targets, labor policies, wages, product lines and the like. Lackawanna's original plant manager, Henry Wehrum, once described as Buffalo's equivalent to Charles Schwab, the founder of Bethlehem, was fired by the Board not for unproductivity but for failing to delegate authority — despite his reputation as a paramount steel producer. Ultimately this separation of owners, directors, and production people would have serious negative consequences on the stability and operations of the Buffalo plant and the company as a whole.

Capital and Profitability

In its early days of operation and on through World War I, Lackawanna Steel was in a stellar position within the steel industry. Well-developed horizontally and well-integrated vertically, it maintained a profitable position even in relation to larger firms. This success was due to what, initially, was a solid financial operation. Stock, while publicly traded, was closely controlled in the first issue. Comparatively few shares, numbering in the thousands rather than millions, were offered at any one time. Large blocks of stock were first offered privately to trusted allies prior to public sale. Second, the stock stabilized at a par value of $100 to avoid speculation and face-value decline. Third, the directors retained a substantial portion of the stock, nearly 50%, for use of the company itself. Thus dividends went straight back to the company treasury to be added to earnings already retained by the company. This money was available for future acquisitions without resort to costly loans and bond issues. By refusing to dilute the numbers of shares and by avoiding large interest payments, the company was able to return rather astonishing dividends in its early years. In 1902 they paid an amazing 100% on equity while in 1909 they paid 50% plus a 15% bonus.

Small stock issues were coupled with restrictions on voting participation; by concentrating controlling interest in the hands of its largest director-owners, the company was avoiding internal squabbles over policy and the shifting coalitions which come with the struggle for control. The three largest owners were the estate of Moses Taylor; the company's chief banker, Farmers Loan and Trust; and New York Life Insurance, the largest non-director institutional investor.

Tight control of the parent company contributed as well to the stability of the subsidiaries. Whoever controls the parent automatically controls the majority-owned subsidiary operations, so there was relatively little opportunity for outside interests to destabilize these operations either.

By 1922 the company had extended its subsidiary control even further so that most of the operations which were only partially owned in 1903 were wholly-owned by the early 1920s. However, later acquisitions were harder to obtain, were perhaps more expensive and a greater drain on capital, so Lackawanna Steel settled for lesser shares simply to have access to beneficial pricing. With a one-third share in Hoyt Mining, a 50% partnership in Negaunee Mine Company, and only 15% of Plymouth Mines, these later acquisitions may well indicate some of the capital weaknesses Lackawanna began to encounter toward the close of its decades of independent operations.

Debt

As was typical corporate practice, Lackawanna raised funds for expansion and acquisitions not only through retained earnings but through borrowing. Typically the steel company borrowed large sums from commercial banks, then sold their own debt in the form of bonds on the open market. The company would pay a redemption to bond holders, giving them interest over and above the face value of the bond, or sell the bond below its face value with promises to pay that value at redemption time. Large funded debts were generally retired through a series of bond issues, with a portion of the debt being redeemed over a period of years until it was paid off. Thus interest would be paid to both bond holders and to the commercial bank for the unpaid portion of the debt.

Lackawanna originally maintained as conservative a position with regard to debt as it did to stock dispersal. In 1903 none of their subsidiaries had any funded debt at all, making their profits free and clear. The company's original $20 million mortgage covered the Buffalo site development, subsidiary acquisition, and operating expenses for a period of 7 years. Furthermore, even as the company acquired increased debts, they were able to generate enough retained revenue to buy back and retire large portions of that debt, thereby reducing interest payments.

Problems began when Lackawanna shifted its emphasis from investing in plant modernization to indulging in further subsidiary acquisition. Some of the growing debt acquired 1910-1922 was translated into actual productivity increases in Buffalo; in 1905 the directors reported that the company's first priority in capital utilization was to add to plant capacity. After that time, however, modernization received less emphasis. Although the 1910 indebtedness soared to $39 million, it was chiefly directed toward greater backward-vertical acquisitions. Existing subsidiaries' stock, once debt-free, became used as collateral, making it less and less possible to retire obligations. By 1922 the company still owed nearly $22 million, a debt worth more than 50% of its stock value.

The most serious consequence of increasing indebtedness and the direction of debt utilization was the trade-off of raw materials wealth against the deterioration of their plant's productive capacity. Even as the company continued to build some new facilities at Lackawanna, they directed little or no money into maintenance of what had been built earlier. When Bethlehem first inspected the New York facility, the Buffalo plant "was found to be antiquated and in such a rundown condition that it was only a matter of time before the Lackawanna Steel Company would have been forced out of business in competition with the more modern plants producing the same class of products."

Corporate Practices

As the fourth-largest steel producer in the U.S., the company participated in industry-wide associations for iron and steel manufacturers. Ostensibly "trade associations", these organizations often served as instruments by which the participating members agreed on prices, market areas, and national policy objectives. Among these associations were the American Iron and Steel Institute and the various product "pools" which "supervised" the production goals of specific commodities such as rails. Lackawanna directors also attended the famous dinners held by U.S. Steel head Judge Elbert Gary. Begun in 1906 these dinners were claimed to be mere social occasions but were complete with formal speakers and discussions of intra-industry policy and strategy. Lackawanna President E.A.S. Clarke was the featured speaker at the May, 1912 dinner in which he presented policy recommendations covering both wages and prices. In this way the industry leaders, Lackawanna among them, sought to counter federal anti-trust regulations and thus overwhelm smaller, more competitive firms.

Lackawanna was therefore clearly engaged in "administered pricing". Prices were set through collusion and not through any form of market force of supply and demand. In times of serious depression, prices did, indeed, fall, but not as individual corporate acts. Instead, companies banded together to set lower as well as higher prices. Companies were urged to avoid what Charles Schwab called "a steel war" which he claimed was induced through competition, and to preserve "a fair price . . . giving a fair profit."

The predominant ways by which early 20th-century steel manufacturers sought to increase profitability lay not in creating demand by cutting prices to attract customers but, rather, in cutting costs and also by increasing market share through mergers.

The percentage of labor costs as part of sales income is not easily calculated for the early 20th century; such figures are not reliable and the evidence largely anecdotal. However, Judge Gary declared that in 1913 rail prices at U.S. Steel were $28-30 per ton while the cost of "direct labor" was $15.13, or 54% to 50.4% of that sale price. He declared that by 1919, the year of the general steel strike, the labor cost per ton had risen to $34.61 while the selling price rose only $17 to $45-47 per ton. This made the labor cost between 76.9% and 73.6% of the total product value. Earlier, in 1909, a census bulletin declared that industry-wide labor costs were approximately 27% of gross product value. The truth probably lay somewhere between these two reports.

Lackawanna and other steel corporations saw labor costs as fair game. Operating in an environment of very limited craft union influence, company profit policies were predicated on aggressive wage cuts and absolute authority over the terms and conditions of work. The company exercised control over employment as a means of increasing profits without obstruction, so that layoffs and shutdowns were accomplished with impunity whenever the company decreed.

The other tactic designed to increase profitability was mergers. In consonance with the industry-wide practice of coordinated production and administered pricing, steel companies appeared largely willing to relinquish their independence in order to achieve economies of scale.

Lackawanna's first attempted joint venture was tried in 1916. As early as September, 1915 rumors of a proposed merger had sent Lackawanna's stock soaring on the market, up 9 points in one day. Simultaneously, the Department of Justice began investigating possible violations of the anti-trust laws. Regardless, by 1916 the Replogle-Donner steel syndicate disclosed that they and several leading banking firms were securing large blocks of stock in the Cambria Steel Company preparatory to a merger of Cambria, Youngstown Sheet & Tube, and Lackawanna. Clearly this was a serious and desired venture since the banks involved were already tied to the Lackawanna Board, with National City leading the way.

The deal, however, was quickly abandoned, and Lackawanna was the cause. Their 4th-quarter profits had suddenly soared, and the directors decided that the stock was worth more than the Replogle-Donner offer of $80 per share. However, Lackawanna had a serious flaw the directors chose to ignore; the combined debt of all 3 companies was $42 million of which $37 million was from Lackawanna. Thus the purchasing syndicate chose to reject Lackawanna's demand for a higher stock purchase price in light of its debt burden.

Despite the fact that 1916 was a good year, Lackawanna's next move was divestiture. In December they began selling some of their now-expensive properties; they sold portions of their coal lands to Charles Schwab, receiving the property's value, less the $1.8 million debt on the once debt-free holding. Later in 1917 they transferred part of their ownership in the Scranton ore properties on the Mesabi Range to another mining operation. The reason for these sudden sales of original raw materials subsidiaries was "the directors' policy to dispose of any properties or assets not necessary or advantageous in the conduct of the company's business." Why coal and ore lands were "not necessary" is unclear, but what is clear was the growing need to diminish the debt load in order to improve profits and allow payment of stock dividends which the company was now issuing only sporadically.

The company's overall profitability was good but unreliable. Between 1906 and 1921, the company had only 3 really poor years: 1908, 1911, and 1921. In these years the company actually lost money; otherwise the firm maintained steady rates of return, usually of 5-8% annually. In those years where rates were down, the causes were the need to pay off an outstanding loan (1909), general industry-wide depression (1908), or industry pressure to lower prices (1911). Its profits soared in times of unusual demand, notably the war years, 1914-1918, then fell along with other steel company earnings in the post-war slump. However, they survived what could have been a serious period of loss with the steel strike of 1919-1920, only to go under just before the economy picked up. It was this delayed post-war slump which precipitated the sale of Lackawanna to Bethlehem Steel. The net earnings decline of the top four companies 1920-1921 was:

United States Steel	66.66%
Bethlehem Steel	29.00%
Midvale Steel & Ordnance	143.00%
Lackawanna Steel	180.00%

The rush toward merger, first in 1916 then in 1922, clearly was not impelled by long-term economic trends; the first maneuver was attempted in a time of profitability, the second as a result of short-term losses amid signs of general recovery in the industry. The Board of Directors' willingness to relinquish Lackawanna as an independent corporation stems from a basic fact: these directors were not steelmen. The founders and all of the subsequent directors were capitalists, pure and simple. Making steel was a means of increasing returns on their investments, not an end. With their myriad corporate and banking connections, their focus was on preserving profitability rather than production. The geographic separation of the Board and top management from the day-to-day operations further reinforced the perception that steel mills and ore mines were merely pawns in the game of profitability. The Board had no vested interest in the site, the mill, or the industry but only in the value of the stock which was increased by selling the company rather than by producing steel. Although several directors such as Moses Taylor had been

LACKAWANNA STEEL PLANT BUFFALO 1903

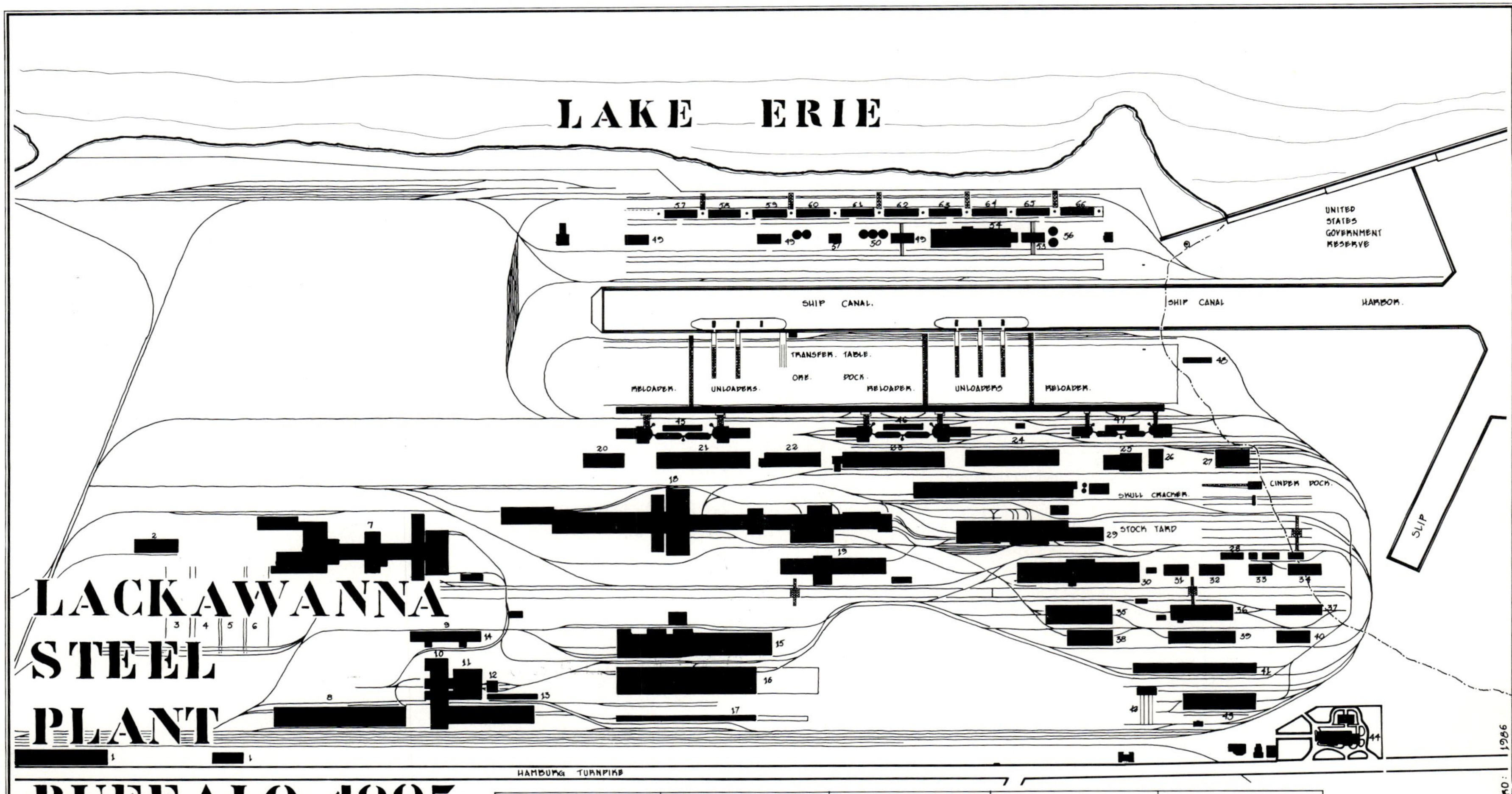

1 BRICK SHED.
2 SPLICED BAR FINISH.
3 STORAGE FOR BEAMS.
4 STORAGE FOR CHANNELS.
5 STORAGE FOR ANGLES.
6 STORAGE FOR Z BARS.
7 RAIL MILLS # 2.
8 HOT BEDS.
9 BOILER HOUSE # 4.
10 48" UNIVERSAL PLATE MILL
11 PIT FURNACES.
12 STRIPPER.
13 GAS FURNACES.
14 BOILER HOUSE.
15 8 - 12 - 16 MERCHANT MILL.
16 OPEN HEARTH FURNACE.
17 GAS PRODUCERS.
18 RAIL MILL # 1.
19 BLOOMING MILL.
20 BOILER HOUSE #3.
21 BLOWING ENGINE HOUSE #3.
22 PUMP HOUSE.
23 BOILER HOUSE #2.
24 BLOWING ENGINE HOUSE # 2.
25 BLOWING ENGINE HOUSE # 1.
26 BOILER HOUSE # 1.
27 LADLE HOUSE.
28 POWER HOUSE.
29 STEEL WORKS.
30 FOUNDRY.
31 PATTERN STORAGE.
32 PATTERN SHOP.
33 PATTERN LUMBER.
34 CARPENTER SHOP.
35 MACHINE SHOP.
36 PLATE SHOP.
37 DUPLICATE MACHINE.
38 FORGE SHOP.
39 PIPE SHOP TOOL HOUSE.
40 STORE HOUSE.
41 ROLL SHOP.
42 CASTING MACHINE.
43 BRIDGE SHOP # 1.
44 GENERAL OFFICE.
45 BLAST FURNACES #6 #5.
46 BLAST FURNACES #4 #3.
47 BLAST FURNACES #2 #1.
48 OFFICE
49 COAL BIN.
50 AMMONIA LIQUOR.
51 BOILER.
52 AMMONIA LIQUOR.
53 COAL BIN.
54 CONDENSING AMMONIA HOUSE.
55 COAL BIN.
56 TAR TANKS.
57 COKE OVEN # 19.
58 COKE OVEN # 17.
59 COKE OVEN # 15.
60 COKE OVEN # 13.
61 COKE OVEN # 11.
62 COKE OVEN # 9.
63 COKE OVEN # 7.
64 COKE OVEN. # 5.
65 COKE OVEN. # 3.
66 COKE OVEN. # 1.

DELINEATED BY: FRANCISCO · HIDALGO-ROMERO: 1986

HISTORIC AMERICAN ENGINEERING RECORD
NATIONAL PARK SERVICE
UNITED STATES DEPARTMENT OF THE INTERIOR

LACKAWANNA · STEEL · PLANT · BUFFALO · 1903
LACKAWANNA, NEW YORK.

SHEET 1 of 4a

HISTORIC AMERICAN ENGINEERING RECORD

involved in the original conception and construction of the Lackawanna plant, he and the others were not true entrepreneurs building brick-by-brick and with the sweat of one's own brow. Using other people's money as well as their own, they valued property in the form of stocks, not steel-making facilities. The move to sell was therefore entirely logical. On 9 October 1922 they closed their books and turned over their assets to Charles Schwab's massive steel empire, thus ending over two decades of corporate history and ushering in a new era.

2. THE PLANT

The Lackawanna Steel Company captured the imagination of the industry by transplanting its operations from Scranton to the Lake Erie shoreline south of Buffalo. After operations at the massive new works had gotten underway, the trade periodical, *Iron Age*, assessed the significance of the project:

> Perhaps no other piece of construction work in connection with the iron industry has attracted attention so widely, and various phases of the enterprise, from the first breaking of ground at South Buffalo, have probably been the subject of more comment in steel engineering circles than any other undertaking in the history of the industry.

The Lackawanna project represented the first attempt by a U.S. corporation to build a fully-integrated iron- and steelmaking plant while simultaneously abandoning an inland location in favor of the growing markets and raw material delivery systems around the Great Lakes. As one company representative prophesied,

> The day for the making of steel in inland cities is past. It takes both water and rail facilities and Buffalo has both. All the steelmen know that and some day most of them will have interests here.

While this prediction was not realized to its fullest extent in the Buffalo area, other centers of steel production around the rim of the lakes would follow the trend set by Lackawanna.

Company engineers were provided with an across-the-board opportunity to install the best contemporary technology on an undeveloped site. The 1500-acre tract at their disposal was noteworthy in itself; Homestead, then the largest plant in the nation in terms of steelmaking capacity, occupied only 156 acres. Construction at Lackawanna commenced in the summer of 1900. The earliest efforts involved building a series of support shops which by themselves constituted ". . . a far greater plant than may be seen anywhere else in or near Buffalo."

The layout of the works centered on a ship canal for receiving iron ore and limestone via lake vessels. The site configuration placed the coke ovens to the west of the canal; the ore yards, blast furnaces, Bessemer and open hearth steel departments, rolling mills, and shops were situated to the east. All the equipment necessary for the various stages of iron and steel production was aligned in a series of parallel rows conforming to the flow of materials through the plant.

Lackawanna's innovative machinery included technology on the cutting edge of early 20th-century engineering: the recycling of waste gas from the blast furnaces for power production and ironmaking operations; by-product coking on the regenerative principle which economized on heat and fuel while capturing volatile gases for subsequent use; the widespread application of electricity. These interlocking heat and energy networks became the basis of the plant's circulatory system while intramural railways and cranes served as its muscles for moving heavy loads of nourishing raw materials, products, and waste.

However, state-of-the-art technology was not the full story. The latest innovations co-existed with up-dated versions of older designs and some surviving techniques from an earlier era. At Lackawanna Steel's chosen new site, this uneven mix-

Plan of the Lackawanna Steel Co. plant as of 1903.

ture of production equipment established the physical environment that a new generation of steel workers would confront.

Coke Ovens

The original equipment for making coke included mechanized coal handling apparatus and by-product ovens equipped with combination charging and pushing machines. Bituminous coal from the company's properties in Indiana and Cambria counties, Pennsylvania, arrived by rail. An 80-ton capacity Hullett car dumper emptied each railroad car; the cars were clamped onto a pivoting cradle and then inverted. The car dumper handled upwards of 30 cars per hour. Belt conveyors then elevated the coal into storage bins.

To prepare the coal for the ovens, it was first dropped from storage into a charging box and then tamped down with electric ramming machines. The charging box was subsequently hauled to the machine that would place the compressed cake of coal inside the proper oven.

Lackawanna Steel's initial plans for the coke plant called for 940 by-product ovens, laid out in two parallel rows of 10 batteries each with 47 ovens per battery. Construction commenced in April, 1902, but only half the projected number of ovens was ever built. By 1904 6 batteries were in operation; there were 10 by 1907, or a total of 470 ovens. The first 4 batteries (188 ovens) used the Otto-Hoffman regenerative heating design with hot stoves and a tall waste heat stack that provided draft to pull gases through the system. The remaining 6 batteries featured Rothberg ovens. The ovens measured 16' wide x 6½' high x 33' long.

View looking northeast from coke ovens across ship canal to Hullett unloaders, ore yards, reloading bridges, and blast furnaces (c.1905).

Producing coke required baking the charge in the heated retort oven for 30 hours. The high sulphur content of coal from the company's mines soon proved troublesome; despite washing and other coal preparation measures, substandard coke affected the quality of the plant's iron and steel production. Contrary to its original plans, Lackawanna Steel was eventually compelled to begin purchasing coke rather than relying exclusively on its own resources.

A combination charging and pushing machine moved material in and out of the ovens; coal was introduced through a door opening on one side, and coke was expelled out the opposite side. During the charging phase a long peel mounted on a rack-and-pinion mechanism inserted the cake of coal. At the end of the coking cycle an electrically-powered ram pushed the coke out of the oven and into a pan. There a workman quenched the coke with a water spray to halt combustion. A gantry crane with a 10-ton hoist at each end then picked up the coke pan and emptied it into railroad cars by tilting one end. By 1907 charging of crushed coal through the top of the ovens had been substituted for side charging of tamped coal.

The original by-product plant recovered ammonia, tar, and other materials from gases and vapors given off during the coking process. As war in Europe began stoking demand for munitions, an Otto benzol plant was added in 1915 to recover various light oils including toluol; toluene, a toluol derivative, was used for manufacture of high explosives.

As of 1918 Lackawanna Steel began constructing a new coke battery with 60 Semet-Solvay ovens. This design conserved heat and fuel through a system different from the Otto-Hoffman and Rothberg ovens. A pair of regenerative chambers was located beneath the ends of the alternating coking and heating chambers that made up the battery. The regenerators contained a checkerwork of refractory brick that alternately reheated air prior to combustion or absorbed heat from the passage of hot waste gases given off by the combustion process taking place above. Combustion occurred in horizontal flues as heated air from the regenerators mixed with fuel gas. The fuel was admitted at one end of the flue while waste gas was drawn off from the other end. The flow of fuel and waste gas was periodically reversed so the regenerative chambers could alternate between absorbing heat or preheating air.

COKE OVENS
Hand quenching of coke in pan, a practice that continued into the 1920s (1904).

The Semet-Solvay battery was also the first at the plant to quench coke in a hot car. The coke was pushed from the oven into a railroad car and carried to a

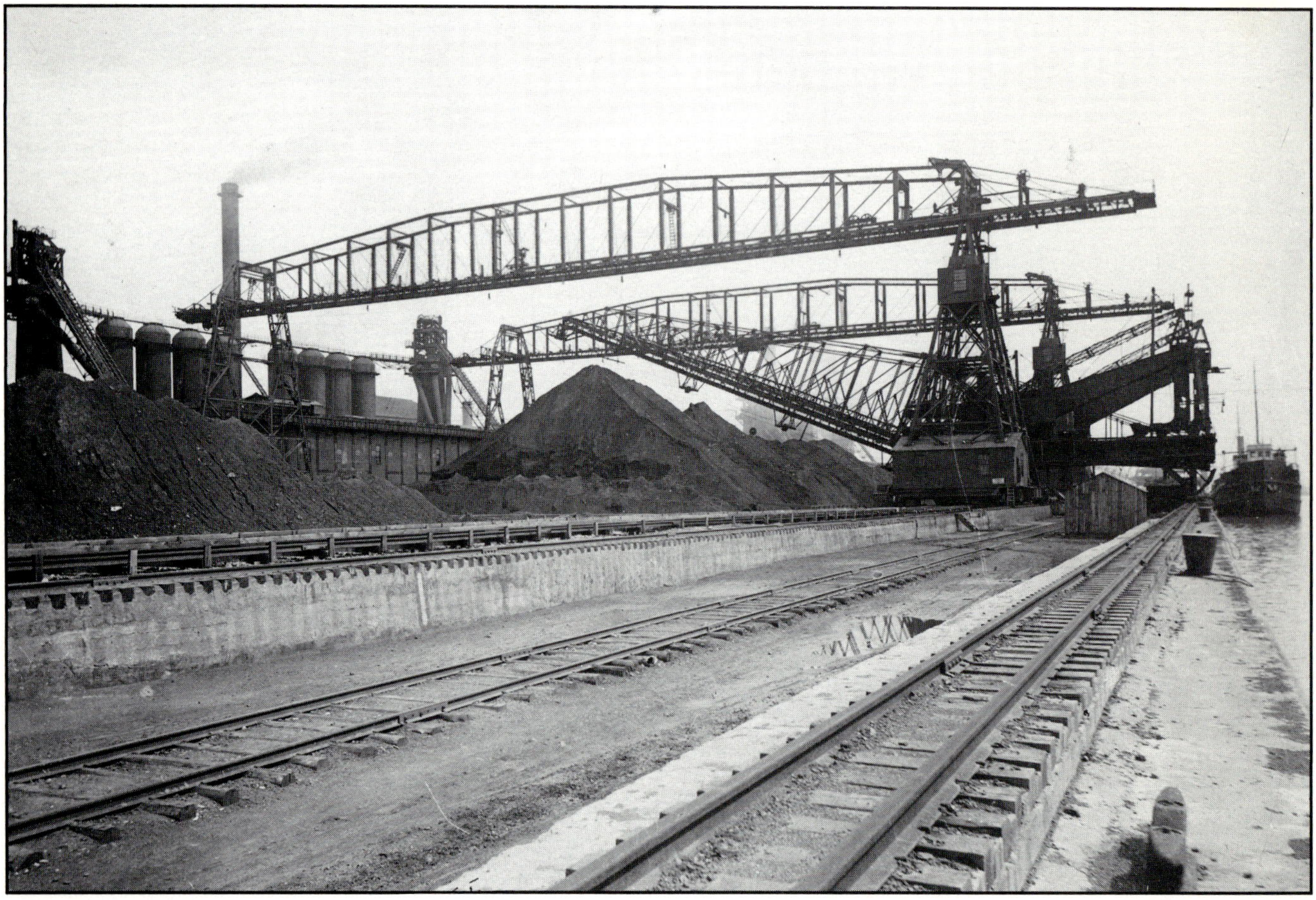

quenching station, there water sprays performed the work formerly done by a man with a hose. The Semet-Solvay battery pushed its first coke in 1920 as three of the original batteries were shut down for dismantling.

Blast Furnaces

Lackawanna Steel's blast furnace department combined contemporary advances in fuel economy and materials handling with new departures in power generation and energy recycling. This mixture of engineering innovations was distributed throughout the sequence of operations necessary for making pig iron: unloading ore and limestone from lake vessels; reloading raw materials from storage yards at dockside into bins near the furnaces; charging the ore, limestone, and coke into the furnaces; smelting the ore; handling the iron as molten metal or solid pigs.

Lake vessels docked along the south side of the ship canal. Hullett unloading machines travelled along the dock and were positioned to remove the majority of the bulk cargo. Each of the 5 Hulletts resembled a giant praying mantis on wheels, including a vertical leg that descended into the hold of the ship. A bucket at the bottom of the leg could grab 10 tons of ore at a time, and the battery of machines was capable of emptying a ship at a minimum rate of 200 tons per hour. Hydraulic cylinders furnished power for operating the bucket leg which could be rotated to reach more of the hold than was possible with a rigid arm. The Hulletts were designed to remove 90-95% of the cargo. The remainder was evacuated by men with shovels and hoisting techniques derived from the early days of ore shipments on the lakes.

The Hulletts deposited their bites of ore onto storage piles at the rear of the machine or into hopper cars drawn up underneath. Three reloading bridges, driven by electricity, were erected to transfer materials from ground storage to the stocking trestle and its system of bins. Each reloader included a 7½-ton bucket for lifting materials out of the stockpiles.

COKE OVENS
20-ton gantry transfers coke from quenching pan to railroad cars (c.1904).

Bins for holding ore and limestone prior to charging stood parallel to the ship canal, between the ore handling equipment and the blast furnaces themselves. A trestle connected the bins served each furnace. Dump cars on the trestle received material from the reloading bridges and distributed it to the proper receptacles. Electric scale larries were filled from the bins overhead; the larries weighed each load of ore or limestone before delivering it to the skip cars. When the plant first began operating in 1903, coke was not stored in the bins with the other raw materials. Instead, it was transported over from the ovens in railroad cars with tilting pans and then transferred to the skip cars. However, Lackawanna abandoned this procedure by 1907.

Charging skips running up inclined bridges carried alternating loads of ore, coke, and limestone to the top of the furnaces. The top construction featured the double bell and hopper system for charging the finer Mesabi ores as well as reducing the amount of waste gas that escaped into the atmosphere. A movable oval deflector, suspended from chains, distributed each skip car load into proper position on the small bell below. Gases taken off from the top passed through a three-branched downcomer into a dust-catcher. After further washing, the waste gas was recycled for heating the hot-blast stoves and driving the engines that blew air into the furnaces.

Lackawanna's blast furnace installation initially consisted of 7 stacks. Numbers 1 and 2 had been dismantled in Scranton and reconstructed at Buffalo. Each stood 87' high and measured 17' in diameter at the bosh, the widest part and zone of highest temperature. Beside each furnace were four central combustion hot-blast stoves, 85' high x 18' diameter, also transplanted from Scranton. These stoves represented another example of heating on the regenerative principle. Each stove consisted of a combustion chamber where gas was burned to heat regenerative flues of brick checkerwork. These heating elements were enclosed by a brick-walled cylinder and a steel shell. As at the coke ovens, the brickwork in the stoves first absorbed heat from passing waste gases and then transmitted the retained heat as air was introduced.

BLAST FURNACES
Ore docks. Hullett unloaders and reloading bridges transfer ore or limestone from vessels in ship canal (R) to stocking trestle (c.1908).

Blowing engines propelled air into alternating stoves where the blast was heated to 900° in passing through the flues. Located in Blowing Engine House #1, the equipment for #1 furnace consisted of three horizontal reciprocating engines powered by steam raised in boilers heated by waste gas. These Dickson steam blowing engines had also seen service at Scranton; one had been built as early as 1879. They delivered a relatively small amount of air per revolution: 560 cubic feet at 15 psi, a pressure level associated with the practice of fast driving. By contrast, the two I.P. Morris beam engines that serviced #2 furnace furnished 940 cubic feet of air per revolution. The comparatively low temperature and volume of the air blast limited the nominal output of these two furnaces to less than 300 tons per day; reconstructed at the new plant site during 1901-1902 and blown in on 11 February 1903, #1 and #2 furnaces were actually producing 260-285 tons of pig iron over 24 hours of continuous operation late in that year.

Lackawanna's other 5 blast furnaces featured larger stoves and more powerful blowing engines. The furnaces themselves were 94' tall — a reduction from the planned height of 110' based on the prevailing opinion that extra height did not necessarily lead to greater output. Stoves for #3-#6 furnaces measured 133' x 22'; the dimensions of #7's stoves were 109' x 22'. Located north of #1 and #2, #7 was equipped with a McKee stock distributor instead of the movable oval deflector used on the plant's other furnaces. An electric motor rotated the McKee distributor after each charge to minimize segregation of coarser and finer materials that could lead to temperature imbalances inside the furnace.

BLAST FURNACES
Foreground: stocking trestle with hopper cars. (L) four regenerative hot-blast stoves. (R) furnace with inclined skip bridge for charging raw materials.

The blowing engines for the larger blast furnaces represented one of the principal engineering innovations at the new Lackawanna plant. During the 19th century American furnacemasters had borrowed the European practice of capturing some of the waste gases as boiler fuel for making steam. By the outset of the 20th century, energy-conscious Europeans had also demonstrated that it was feasible to recycle waste gas for other forms of power production such as generating electricity. Henry Wehrum, Lackawanna's first general manager and chief designer, planned to make extensive use of these latest techniques for reducing waste and economizing on fuel. As he put it, "We will get our power from the tops of the stacks. The smoke will be our power."

Blowing Engine House #2 contained the first installation of this novel equipment for supplying air blast to the iron furnaces: horizontal double-cylinder gas engines driving vertical blowing engines. The De la Vergne Refrigerating Machine Company of New York built the gas engines according to patents held by Ernest Koerting of Hanover, Germany. Rated at 2000 hp each, these units delivered three times as much air as the older blowing engines for #1 and #2 furnaces. On 8 June 1904 #3 furnace began operating on gas engine power with projected cost savings of 3:1 over the steam engines previously in use, the relative cheapness of coal and the expense of gas-cleaning notwithstanding.

Combined with a higher blast temperature of 1000°, greater blowing capacity created expectations of significantly higher output. The daily capacity of the larger furnaces was originally projected at 800 tons. However, when #3-#7 furnaces were blown in during 1904-1907, actual production averaged 500 tons, equivalent to the record output for which the Carnegie furnaces at Duquesne had been designed in the late 1890s.

The gas engines used for electrical generation likewise failed to realize the great expectations of their engineers. Power House #1 contained eight 1000 hp Koerting engines connected with apparatus for producing both direct and alternating current. Unfortunately, the engines proved highly sensitive to variations in the chemical composition of the waste furnace gas. Moisture leaks around the tuyeres that delivered the air blast raised the hydrogen level in the waste gases; fluctuations in the mixture fed into the engines resulted in greater speed variations than were suitable for generating electricity. European experts on gas engine operation and American electrical engineers scratched their collective noggins over this problem. However, the eventual solution turned out to be greater dependence on electric power transmitted from Niagara Falls than Wehrum had thought would be necessary.

BLAST FURNACES
2000 hp Koerting horizontal gas engines for driving vertical blowing engines that supplied air to blast furnaces (c.1905).

While pig iron production at Lackawanna at first fell short of the company's most optimistic projections, there was still plenty of metal and slag to haul away. Most of the molten metal was used directly for steelmaking. However, to provide for metal made on Sundays or other periods when the Bessemer shop was idle, two Heyl & Patterson pig-casting machines were installed. This equipment superseded casting the pigs in sand beds and then removing them manually.

Major additions to the blast furnace department occurred in 1918 when #8 and #9 furnaces were blown in. Located at the northern end of the plant's blast furnace row, the new stacks each measured 94' x 23' with four 110' x 22' stoves. As of 1920 Lackawanna's annual pig iron capacity was 1,440,000 tons from 9 furnaces, up from 1,080,000 as of 1908. Other equipment as of 1920 included a total of 6 vertical steam blowing engines and 21 gas blowing engines as well as 5 double-strand Heyl & Patterson pig casting machines.

Steelmaking

The Scranton plants of the Lackawanna Iron & Steel Company had made steel with Bessemer converters alone. Plans for the new lakeshore site envisioned the same concentration on Bessemer rail manufacture. However, during the first phase of construction Lackawanna Steel decided to add open hearth capacity and broaden the line of products it rolled. Thus, the plant came to contain updated versions of two very different steelmaking techniques which had originated in the mid-19th century.

BLAST FURNACES
Cast house floor. (L) jib apparatus for mud gun used to close tap hole. (C-R) molten iron in runner; splasher to protect furnace above tap hole from hot metal; skimmer to regulate level of metal flowing from furnace (1910).

The layout of the steelworks and rolling mills at Lackawanna embodied a traditional conception that certain types of steel were preferable for certain products: acid Bessemer for rails, the staple line of 19th-century mills; basic open hearth for structurals or plates. In other words, the plant's design represented a transitional moment in the history of metallurgy as open hearth steelmaking began to overtake the once-dominant Bessemer process; Lackawanna's engineers froze that moment on the ground as they laid out their steelmaking capacity in relation to certain rolling mills. By contrast, the Tennessee Coal & Iron Company's plant at Ensley, Alabama was the first to break the traditional association of acid Bessemer steel with rail manufacture in 1900-1901; by 1908 yearly open hearth production had surpassed Bessemer tonnage for the first time in America, and U.S. Steel's enormous Gary works had been designed to roll open hearth rails. Thus obsolescence threatened the Bessemer rail even as Lackawanna was being constructed.

The Bessemer steel plant at Lackawanna was built to use both direct metal from the blast furnaces and cold pig remelted in cupolas. To accommodate the mechanical requirements of these differing procedures, the Bessemer building included both a cupola house and a mixer house where molten pig carried in 25-ton ladles was poured into a 300-ton metal mixer. First devised by Captain Bill Jones of Carnegie's Edgar Thomson plant at Braddock, Pennsylvania, the mixer acted as a reservoir that conserved heat, blended the products of several blast furnaces, and supplied hot metal of relatively uniform composition for steelmaking.

On a line to the south of the cupola and mixer houses lay the converter house proper, containing 4 vessels rated at 10 tons each. The air which initiated the conversion of iron into steel was forced into the converters by 4 steam-powered blowing engines located in Power House #1. Capable of furnishing air at 24-26 psi — the upper limit of contemporary practice — these engines provided enough capacity for blowing two heats at once, each blow lasting about 7 minutes.

Ancillary facilities in the Bessemer works included speigel cupolas and a furnace for melting ferromanganese which was added to the steel in the ladle as a means of replacing the carbon expelled by the air blast and otherwise improving the quality of the heat. The ladle house also contained equipment for making and drying converter bottoms, which had to be replaced fairly frequently, and likewise the stoppers for the ladles from which the steel was teemed into ingot molds.

STEELMAKING
Teeming ingots at pouring platform in Bessemer converter house. A 40-ton hydraulic jib crane holds the 12-ton ladle over the line of ingot molds on narrow-gauge cars (1910).

Steel from Lackawanna's Bessemer plant was used primarily in the heavy and light rail mills. The first Bessemer blow took place on 20 October 1903. The annual

capacity of the works was eventually rated at 845,000 tons, somewhat less than the original projections of a million tons or more per year.

Making steel in the early years of the 20th century was a matter of linking improved metallurgical science with advances in materials handling efficiency. Not all the links were joined at once. Since the Lackawanna plant was built during a transitional period, it used a mixture of techniques for moving materials. For example, in the Bessemer converter department modern overhead electric travelling cranes with DC motors contrasted with older revolving hydraulic jib cranes for pouring the steel into the ingot molds and hand-operated cranes for replacing converter bottoms. Still, when heavy materials travelled through the air or along the ground, it was usually with some form of mechanical assistance, marking the continuation of engineering developments that had first begun reshaping steel plant technology and work after the Civil War.

Lackawanna's addition of open hearth steelmaking to its original plans resulted from the company's decision to augment its product line with heavier structurals and plates; #1 open hearth shop was sited close to those particular rolling mills. A total of 14 basic open hearth furnaces was built in #1 shop between 1903 and 1912 with six 60-ton units installed in 1903-1904; five 75-ton units in 1905-1906; one 75-ton unit in 1909-1910, and two 75-ton units in 1911-1912. A 600-ton metal mixer was also completed during 1911-1912. Laughlin or Morgan gas producers provided fuel for the furnaces. Number 1 open hearth made its first steel on 26 September 1904.

STEELMAKING
Charging machine, #1 open hearth. This apparatus replaced manual labor for the heavy work of loading cold materials such as scrap, ore, and limestone in the furnaces (1904).

In 1912 Lackawanna Steel undertook construction on a second open hearth department, located directly north of the Bessemer works. The first two furnaces in the new shop were 200-ton tilting models, designed by company engineers for use in conjunction with the nearby Bessemer plant. Combining acid Bessemer and basic open hearth operations was known as the duplex process, another innovation imported from Europe. It divided the making of steel from pig iron into two phases: removing silicon and carbon via the Bessemer blow then removing phosphorous and sulphur with additional refining and finishing in the open hearth. This hybrid method attempted to combine the speed of the Bessemer process with the quality control possible in the open hearth. Compared to conventional stationary open hearths, tilting furnaces further reduced the time required for steelmaking; oscillated back and forth for charging and tapping, they were designed to retain a reservoir of molten metal that accelerated the necessary chemical reactions.

The first duplex heat in #2 shop was tapped on 29 May 1913. For the 12 months from April, 1915 through March, 1916, Lackawanna Steel made a total of 453,619 tons by duplexing — more than a quarter of the plant's total output. Between 1913 and 1916, the heats from the tilting furnaces averaged 102.7 tons. The company also placed 8 stationary furnaces in #2 shop. Four 100-ton basic furnaces went into operation in 1915, three more in 1916, and the last unit in 1917.

Over a period of less than two decades, the focus of steelmaking at Lackawanna — as elsewhere in the industry — thus shifted from the Bessemer process to the open hearth. The plant's annual ingot capacity escalated from 845,000 tons in 1903 to 1,345,000 tons in 1908, 1,600,000 tons as of 1916 and 1,840,000 tons as of 1920. Most of this growth was due to the proliferation of open hearth furnaces. Five decades earlier, the pyrotechnics of the Bessemer converters had heralded the dawn of a new age in American steel production. But most of the workers who made steel during the first half of the 20th century would find that their jobs were shaped by the technology and work of charging, operating, tapping, teeming, and repairing the slower but more versatile open hearths.

Rolling Mills

STEELMAKING
Tapping side of furnaces, #1 open hearth. (L) pouring platform for teeming ingots. (C) 100-ton ladle crane and railway layout. (R) steel ladles, square slag receptacles (1908).

When Lackawanna Steel's founders viewed the plant in late 1906, they pronounced their original plans complete. Seven rolling mills were available for transforming steel ingots into a variety of products. In order of first rolling dates, the

mills and their products were as follows:

#1. 32" rail mill; heavy (60 lbs./yd. or more) rails. 20 October 1903.

#2 and #3. 24" rail and structural mills; one roll train for light rails or splice bars and one for structural shapes. 5 July and 15 September 1904.

#4. 32" slabbing mill; slabs, blooms, and billets. 16 October 1904.

#5. 48" universal plate mill; universal as well as 72" sheared plates. 5 November 1904.

#6. 16"-12"-8" combination merchant mill; bars for railroad car builders, agricultural implements, and boltmakers. 31 January 1905.

#7. 40" blooming mill; billets. 27 November 1905.

The types of rolled products indicated a strong orientation toward capital goods markets in the fields of construction and transportation. Bessemer ingots, for the most part, went into the two rail mills, and those buildings were situated on the same general north-south axis as the converter shop. The slabbing, plate, and merchant mills stood close to #1 open hearth.

In most of the mills cycles of heating alternated with rolling. Enclosures called soaking pits equalized the internal and external temperatures of steel ingots that were subsequently rolled into semi-finished blooms or slabs by the primary mills. Reheating furnaces performed comparable functions of controlling temperature and molecular structure before the blooms or slabs passed through the finishing mills to emerge as structural shapes, plates, or merchant bars.

All the rolling mills were driven by steam engines, even though plant engineers had applied electricity for a number of other novel purposes. Driving the mills with traditional prime movers was yet another indication that steel engineering was passing through a transitional stage during the years when the Lackawanna plant was built. The designs of the mills represented solutions to the problems of rolling various shapes, and those engineering choices created a variety of work processes.

The centerpiece of the Lackawanna plant was the #1 rail mill, a structure a third of a mile long from the stripper house to the finishing department. Consisting of 6 stands with two-high rolls driven by 5 steam reversing engines, the mill was designed to roll rails from 18½" square Bessemer ingots in 15 passes without reheating. In other words, a square ingot was transformed into a T-shaped rail after passing 15 times between 6 pairs of grooved upper and lower rolls, being reversed back and forth in several pairs of rolls.

This method of rolling rails differed from mainstream American practice on two important points: first, by dispensing with intermediate reheating furnaces, and, second, by using reversing mills. Rolling without reheating the semi-finished bloom was inspired by a current theory that improving the durability of Bessemer rails depended on finishing them at relatively low temperatures. Two-high reversing rail mills were more common in Europe than in the U.S. where American engineers had perfected three-high mills. The new Lackawanna mill could trace its ancestry back to Henry Wehrum, a veteran of the steel industry in Alsace-Lorraine, who had previously installed reversing mills at the Scranton Steel Company in 1883. Design and technology thus established job requirements and production processes that differed from contemporary rail plants in some particulars.

At Lackawanna as of 1903 the operation of rolling rails began when railway cars brought congealed ingots from the Bessemer shop to the stripper building. There, two duplex hydraulic strippers removed the molds into which the molten steel had been teemed; four molds were removed simultaneously. The exposed ingots were then carried to the soaking pits, consisting of four vertical furnaces each with four 5' x 7' pits capable of holding six ingots. The soaking pits were fueled by gas producers that burned bituminous coal. Hydraulic cylinders opened and closed the covers over the pits. Two 7½-ton electric overhead cranes deposited the ingots in the soaking pits and extracted them again when their temperatures and physical condition became sufficiently uniform. An electric buggy transferred the ingots to the mill tables for the actual rolling process.

The first three stands of the 1903 rail mill were aligned in tandem, one behind the other on the east side of the building. These stands were known collectively

STEELMAKING
Teeming ingots at pouring platform, #1 open hearth. The stopper rod regulates the flow of steel through the nozzle in the ladle bottom. A 100-ton crane carries the ladle over the ingot molds on waiting cars (1908).

as the blooming mills. Carried on electrically-driven table rollers, the ingot passed once through the first two stands and four times through the third. These six passes reduced the cross-section of the ingot from 18½" x 18½" to 8" x 8". Hydraulic manipulators turned the bloom between passes to expose all its sides to the pressure of the rolls. Hydraulic transfer apparatus shifted the bloom across the mill to the roughing stand on the west side of the building. At that point, a hydraulic shear cropped the front end of the bloom before it passed four times through the reversing rolls. Mechanical manipulators again turned the piece between passes. A steam shear cut off the rear of the bar before a line of rollers carried it forward to the first of two finishing stands. Four passes in the first finisher commenced the final shaping of the rail. A mechanical transfer then moved the bar across to the east side of the mill where the second finisher, the last of the six stands, imparted the 15th and final pass. In normal operation four or five rails were rolled in the various stands simultaneously.

What had started out as a square ingot was now a rail 120-180' long. A line of rollers carried the rail to three electric hot saws for cutting into standard lengths. A cambering machine then curved the rails to correct for distortions during cooling, and a stamping machine branded them with letters and numbers identifying the portion of the ingot from which each rail had been made. The roller line subsequently delivered the rails to four hot beds where they were held for the finishing department. Buggies fastened to wire cables dragged the rails across the hot beds to another series of roller tables and transfers that converged on the finishing shed. Finishing processes included cold straightening under presses and drilling holes for the splice bars that would join the rails into track. After inspection an electrically-powered machine loaded the rails for shipment.

ROLLING MILLS
Electric stripper on crane runway removes molds from open hearth ingots on cars at slabbing mill (1908).

At full production this impressive array of machinery was designed to turn out 600,000 tons of rail annually — greater capacity than U.S. Steel's existing South Chicago and Edgar Thomson works, though the Gary rail mill built a few years later was larger. Lackawanna's rail mill contained some atypical and transitional designs, but its machinery still represented the combination of capital with engineering that had transformed many of the jobs once prevalent in 19th-century rolling mills. However, not all rolling mill operations were mechanized to the same extent. In the #6 bar mill, for example, where steel was caught by hand, production techniques still called for the skills of earlier generations.

During its remaining years as an independent company, Lackawanna Steel added several new rolling mills:

#8. 18" continuous sheet bar and billet mill; operated in conjunction with the 40" mill, supplying semi-finished steel for the hand-operated sheet mills of the nearby Seneca Iron and Steel Company in Blasdell. 15 December 1907.
#9. 10" bar mill; merchant bars and wire rods. 5 September 1911.
#10. 8" bar mill; merchant bars. 5 February 1919 (first proposed in 1915).
#11. 16" bar mill; merchant bars. 25 October 1916.

Lackawanna's expansion of bar mill capacity was prompted by the company's desire to diversify beyond its original emphasis on heavier product lines. Between 1908 and 1920 combined capacity in the rail, structural, and plate mills essentially remained constant. Bar mill capacity tripled from 75,000 tons to nearly 300,000 tons, and these lighter merchant mill products rose to 20.6% of finished output as opposed to 6% initially. The newer merchant mills continued to incorporate incremental changes in technology and motive power as they were built over the years. However, manual rolling techniques and work processes still continued to coexist with methods that required less labor.

ROLLING MILLS
Soaking pits, #1 32" rail mill. A 7½-ton Morgan Engineering Co. vertical bar-type electric charging crane extracts a Bessemer ingot from a pit furnace used to equalize temperatures of steel before rolling (1910).

Between the initial phase of plant construction that began in 1900 and Lackawanna's acquisition by Bethlehem Steel in 1922, production technology combined bold innovations with incremental improvements. However, pockets of older methods still persisted. The plant's equipment represented trends in early 20th-century iron and steel engineering where capital investments were designed to reduce labor, energy, or raw material requirements. This linkage between business priorities,

technology, and work precipitated some decisive breaks with the jobs, skills, and traditions developed by workers in earlier plants. However, technological change did not have a uniform impact on all production and maintenance workers, even at one of the newest steel plants in the industry. The social relationships between the corporation and the steelworkers, on the other hand, added another new dimension to daily life in the plant whose consequences were more comprehensive.

3. WORK

Work and Technology

In 1900 a Buffalo reporter rhapsodizing about the Lackawanna project wrote that, ". . . the steelworker of but ten years ago will feel himself lost when he makes a visit to the colossal plant now being constructed at Stony Point." This journalist's observation contained an important insight as well as a significant oversimplification. Lackawanna Steel and its engineers took advantage of their initial opportunity to install numerous technical innovations that were crystallizing around the turn of the century. Compared with earlier iron or steel mills, capital and engineering had transformed many jobs, creating a virtually new plant environment in their drive for lower production costs. Workers who had previously moved materials by heavy manual labor felt the effects of new technology most directly. Electric overhead cranes and other materials-handling apparatus decimated the ranks of the labor gangs that had once assisted skilled workers.

ROLLING MILLS
Blooming rolls of 32" rail mill reduce ingot in cross-section while increasing elongation as ingot passes from right to left (1910).

However, as had been the case during the iron and steel industry's earlier development, technological change in various sectors reached these new plateaus at different rates. Within the complex world of an integrated steel plant such as Lackawanna, some workers would now operate new or improved machines; others would continue to labor in the shadow of machines at difficult tasks that defied mechanization or on simpler jobs where substituting capital for labor was uneconomical. Also, the effects of technological change were less obvious at the upper skill levels than in situations where laborers' jobs became redundant. Blast furnace keepers, first helpers on the open hearths, and rollers were not displaced; if anything their empirical knowledge and accumulated experience became more indispensable for iron and steel production. It was the redefinition of skilled workers' roles relative to the company and other workers that provided the most significant point of contrast with the experiences of preceding generations.

As a final caution against overgeneralizing about technology and its effects on work, we should remember that at Lackawanna's new plant there were a number of other job settings besides those directly involving iron and steel production. It took a corps of bricklayers to keep all the refractory linings in good repair. New power generation and transmission methods required the skills of electricians and millwrights. Riggers were called upon to solve problems involving materials movement during construction or repair. At the plant's forge, foundry, pattern shop and machine shop, the mix of equipment, jobs, and skills resembled a factory complex making heavy equipment rather than a steel plant. It is easy to overlook maintenance and support shop workers when describing the more spectacular operations involved in making iron or steel. However, the division of labor in an integrated steel plant did consist of many more separate work situations than the following pages mention.

A number of jobs in the blast furnace department at the new Lackawanna plant involved operating machinery to accomplish tasks formerly done by human effort alone. The most recent of these innovations transformed older methods of stocking the furnace with raw materials and handling the iron after it had been reduced to molten pig.

ROLLING MILLS
32" rail mill. Square ingots are reduced in cross-section and elongated while passing from left to right through (R) 1st and 2nd blooming rolls. The bloom returns from right to left for passes through (L) roughing stand (1910).

By the mid-1890s furnaces could churn out pig iron at a rapid pace due to fast-driving with higher blowing pressures and air volumes. However, manual methods

Rush
1479

for supplying the furnaces with raw materials remained a bottleneck. To correct that discrepancy while also slicing labor costs, Carnegie Steel installed mechanical equipment for stocking and charging at its Duquesne Furnaces near Pittsburgh. Lackawanna Steel adopted similar procedures when building its new works a few years later. Electric larry cars — small hopper cars with automatic scales to weigh ore and stone — replaced the jobs of the bottom-fillers at older plants who had used wheelbarrows when moving stock from storage to the furnace. Skip cars running up inclined charging bridges superseded the vertical mechanical lifts that had raised the wheelbarrows to the top of the furnace; skip charging likewise eliminated the top-fillers who had controlled the placement of the raw materials on the furnace bell while working amid clouds of noxious emissions.

During the same period the introduction of the mud gun, invented by Samuel Vaughen of Lorain, Ohio, altered the procedure for closing the tap hole after each cast. At older furnaces the keeper and his helpers had used long poles to stuff balls of clay into the iron notch after the molten iron had drained out. While the men struggled to close the opening, the air blast continued to force sparks, slag, and hot gases out of the furnace. The mud gun made closing the tap hole somewhat less arduous and risky. Operated by the keeper himself, it used compressed air to force a moist mixture of clay and coke dust into the tap hole. As future sociologist Charles Walker was to note during his tenure in a steel mill, using the mud gun "requires a streak of skill." Here was an instance where introducing a machine had displaced manual labor while calling for new abilities of a different sort to operate it effectively.

At integrated plants such as Lackawanna, the practice of taking molten iron directly to the steelmaking departments revised the work of keeper's helpers and sand-bed men. Previously, they had been responsible for constructing the runners through which the iron flowed out of the furnace and the molds in the cast-house floor where it congealed into pigs. Channeling the iron into hot metal ladles for transfer to the steel departments still required maintaining a network of runners (which resembled metal troughs), but sand casting became a thing of the past. The adjustable skimmer, introduced around 1898, also changed the nature of the work in the cast house. Separating unwanted slag from the iron flowing out of the furnace had formerly required building a diversion dam out of sand. Since the height of the dam could not be adjusted to compensate for changes in the level of molten metal, any overflow endangered nearby workers. The adjustable skimmer functioned as an iron regulating gate that could be raised or lowered by hand. Here was a device that was not really a machine but which still had a considerable impact on the work process.

ROLLING MILLS
Rail deseaming machine in 32" mill (c.1914).

When molten iron was cast into pigs for remelting in the steel departments or for sale, the introduction of special machines had replaced casting in sand beds at a number of plants, including Lackawanna. Workers in the pig-bed crew had been responsible for making sand molds in the characteristic form of sows nursing piglets, guiding the iron into the molds using iron bars tipped with slag lumps and removing cold pigs from the beds with sledgehammers and manual hefting. Individual pigs at steel plants could weigh upwards of 100 lbs.

Pig-casting machines devised by E.A. Uehling and other engineers utterly transformed this process. Hot metal now was collected in open-top ladles and poured into a double strand of molds traveling on an endless chain. The pigs cooled in transit as the mold line proceeded and then dropped into railway cars for shipment to their next destination.

At Lackawanna one pig caster attended by 6 men displaced 20 workers from the cast house. The crew on the casting machine now consisted of a foreman, a hot-pot dumper who controlled the discharge of molten metal from the ladle into an intermediate holding trough, a flow regulator who oversaw the transfer of the molten iron from the trough to the slowly-moving molds, a molder and a clay mixer who prepared the receptacles that received the metal, and a laborer.

Pig-casting machines required considerable capital investment, but they did eliminate the heavy jobs of workers who were denigrated as "gorilla men" because their tasks required physical stamina. However, according to one British observer

ROLLING MILLS
Hot saws, 32" rail mill (1910).

the substitution of capital for labor in this process involved more than just economics:

> Furnace men do not claim that the casting machine is more economical than even casting in sand or in stationary chills; but it does away with a very objectionable and troublesome class of labor and it furnishes an iron perfectly free from sand.

The demise of the wheelbarrow men, bottom-fillers, weighmen, top-fillers, and other manual laborers cut the number of men required around the blast furnaces by as much as 40%. At Lackawanna and other advanced plants, blast furnace crews now worked with equipment that would have been unknown a decade earlier. One contemporary analyst who tried to place the relationship between men and machines in historical context felt that workers operating the new electric larries now ". . . must be about as skillful as our trolley motorman." Still, the work environment around a blast furnace remained demanding, the daily routines arduous. The same British observer noted:

> The bulk of the heavy drudgery has been obviated by the use of machinery. There is no pig-lifting, no hand shoveling of stock, no hauling of charging barrows. All the tedious clay work around the hearth, and incessant changing of tuyeres, is done away with. But after all this is done, blast furnace work is and will remain a hard occupation, requiring endurance and muscular strength.

Some of the most demanding jobs now involved maintenance and repair rather than actually making iron, as Charles Walker discovered during his stint on a gang that cleaned the interior of the hot-blast stoves. The situation required special clothing: wooden sandals an inch thick to protect the soles of the feet from residual heat; jackets buttoned to the neck; kersey caps with ear flaps; two pairs of gloves; two bandana handkerchiefs worn over mouth and nose to filter out the pervasive dust; close-fitting goggles. Once bundled up, the men entered the stove by crawling through a door 14" in diameter. Inside, they worked in the combustion chamber, a space 8-10' x 3-4'. The combustion chamber was hot and gaseous, but Walker and his gang remained inside for as long as an hour at a stretch, attacking the cinder with picks, sledgehammers, and hoes. However, they took much shorter turns when poking flue dust out of the holes in the brick honeycomb which was designed for optimal heat retention.

Turn-of-the-century innovations implemented at Lackawanna altered some jobs around the blast furnaces more than others. Mechanized skip charging and pig casting caused dramatic changes in the processes of getting materials into the furnace and handling pig iron, but the spurt of new methods affected the men who tended the hot-blast stoves or disposed of slag much less directly. Blowers supervising groups of furnaces and keepers in charge of individual stacks still needed skills comparable to those of earlier generations. Instrumentation had improved. In the 1850s furnacemen around Youngstown, Ohio had regulated the heat of the blast by relying on the different melting temperatures of lead (612°) and zinc (787°). If a lead strip applied to the cast-iron piping of old-fashioned Thomas hot-blast stoves failed to melt, the blast temperature was too cool; if the zinc strip melted, the blast was too hot. By the turn of the century self-registering pyrometers and recording gauges provided graphic monitoring of variables affecting furnace behavior. But, empirical rules of thumb about the significance of cinder coloration or slag appearance also remained in vogue. Skilled positions still required occasional heavy physical exertion. For example, electric drills lessened the effort necessary to open the tap hole, but the keeper might still be called on to swing a sledgehammer when driving a long pointed bar through the remainder of the clay plug. The jobs of blowers or keepers thus continued to call for a combination of mental and manual abilities during routine and emergency operations. Even technical manuals issued by steel companies recognized that "there are few situations in life where promptness and decision, forethought and good judgment, skill and experience are more needed than about a blast furnace in times of trouble."

At Lackawanna's open hearths the impact of new technology on jobs and work processes was just as significant and uneven as in the blast furnaces. The clearest effects likewise involved new methods of charging raw materials and handling molten products.

Materials such as scrap that were charged cold had been placed in older furnaces by gangs of men wielding long iron bars with ends flattened like a spatula. However, the electric charging machine introduced by Samuel T. Wellman of Cleveland had taken over this exceptionally heavy work. Moving along rails parallel to the open hearths, the long arm of the charging machine picked up boxes from railway buggies and emptied the contents, weighing a ton or more, into the furnace. The operator riding on the machine manipulated six controls and blended the charging arm's eight discrete movements into a smooth sequence. Each charging machine served a group of furnaces and was capable of stocking a 50-ton furnace in 30 minutes. When the Wellman charger was first introduced at the Bethlehem Steel Company it displaced a total of 20 men who had once fed four furnaces by hand.

Two 40-ton travelling electric overhead cranes carried ladles of hot metal from the blast furnaces or, after 1912, from the mixer, to the furnaces in Lackawanna's #1 open hearth shop. A skilled crane operator regulated the transfer of molten iron from the ladle to the furnace. To prevent expensive and dangerous spills, the operator discharged hot metal from the airborne ladle by raising one side via an auxiliary hoist; he tilted the ladle so that iron flowed at the proper rate into a portable runner placed before the furnace.

The method of supplying fuel to the furnaces was likewise mechanized. Adjacent to the #1 open hearth shop, Lackawanna Steel installed a plant of 16 Morgan continuous gas producers for the furnaces built after 1905. These units included variations of contemporary devices, including a mechnical poking arrangement to keep the fires going, that were designed to do away with the laborers who had previously tended gas producers. In addition, Lackawanna's equipment featured the J.R. George automatic coal feeder. This procedure used a distributing crane, operated by a boy, that swept over the entire row of producers rather than charging each unit separately from a less-flexible system of fixed bins. However, hand-fired Laughlin gas producers also remained in use for the 6 original furnaces in #1 shop.

For tapping and teeming steel or cleaning up the pit areas, the introduction of overhead cranes and other materials-handling techniques gradually altered the duties of open hearth crews. The principal modification of earlier practice involved using traveling cranes in conjunction with casting ingots on railway cars and moving them outside the teeming area before removing the molds.

Older plants had relied on stationary jib cranes at each furnace to lift the ladle full of hot steel after the furnace had been tapped. Swinging in an arc, the arm of the jib crane carried the ladle over semicircular rows of ingot molds set in the ground. That arrangement resembled the pit layout in early Bessemer plants. By replacing the older fixed cranes with models that travelled overhead full length of the pouring aisle, steel companies were able to eliminate a number of the helpers formerly required for tapping and teeming. Lifting the molds off ingots cast on the ground had called for heavy and dangerous work by the pit crew. Laboring amid terrific heat from the ingots, they attached chains hanging from a crane to lugs on the molds; if the mold stuck to the coagulating steel, pitmen loosened it with sledgehammers. At newer layouts such as Lackawanna the overhead crane carried a ladle of molten steel to molds standing on railway cars next to an elevated pouring platform. After the molds on each car had been filled, they were hauled away to a mechanical stripper outside the open hearth shop. This practice eliminated the need for men to work around the base of molds in the pit where accidents had been frequent.

The use of overhead cranes for pit clean-up reduced the number of laborers on the tapping side of the furnace in other ways. For example, the steel and slag that spilled into the pit had formerly been carted away by wheelbarrow after it had cooled. Loading slag and scrap into larger boxes that the crane could then pick up required fewer men. Also, the introduction of the slag thimble cut the amount of spillage during tapping. Overflow from the steel ladle formerly had been broken up by laborers or lifted as a solid piece by the crane. Catching most of the overflow

in the thimble illustrated another newer practice, which was not exactly mechanization but which nonetheless reduced the number of laborers in the pit crew and altered the daily work done by those who remained.

Jobs in the pit crews involved labor with both hand tools and machines during the early decades of the 20th century. Charles Walker noted that the knack of working in conjunction with the overhead cranes had become an important litmus test of competence. Within his own pit gang of 8-10 men, figuring out the best way of attaching odd hunks of scrap to the crane provided ". . . one small matter where a grain of judgment and ingenuity has place." On the other hand, dislodging slag with a pick felt more like working in a mine to Walker. Steel companies or other observers might lump all the pit crew's tasks together as unskilled labor, but Walker noted:

> The jobs we did were pick and shovel jobs. But have you ever used a pick on hot slag? There is judgment and knack, and he is a fool who says that 'anyone can do the job.' Whenever the chance for special skill happened by, as in hooking the crane to a difficult piece of scrap, there was an abundance, and much rivalry to show it off.

Steelmaking in Lackawanna's early open hearths required a combination of machines, mental judgment, and muscle — much like the situation over at the blast furnaces, though open hearth work involved more lifting of dead weight. After the steel industry had reorganized work at the open hearths in the wake of technological advances, each furnace was staffed by three attendants: the first helper, second helper, and cinder-pit man or third helper. A foreman or melter oversaw operations at a group of several furnaces.

ROLLING MILLS
24" 3-high finishing stand in #3 structural mill. Hinged tilting tables raised and lowered the steel so it could pass back and forth between the middle and upper or the middle and lower rolls to be shaped (1904).

During this era the first helper had charge of his furnace until the heat was ready to be tapped. He regulated the charging, heating, slag runoff, repairs, and addition of reagents in accordance with the specifications for each heat. His judgment during the working period after the charge had melted was particularly critical. The chemical properties of slag had to be manipulated to promote proper elimination of impurities; the carbon content and temperature of the molten bath had to be kept within tolerances as well. Bethlehem Steel founder Charles Schwab once boasted that he could make a good melter out of a farmhand in less than two months. He must have been willing to tolerate quite a bit of poor steel.

Cooking up the heats was a slow process of blending many ingredients correctly. If a test sample indicated that the carbon content of the bath had fallen too low, hot metal from the mixer was added to the furnace. If the carbon content proved too high, iron oxide, usually in the form of lump ore, was added to promote the necessary chemical reactions. The composition of the slag could be adjusted by introducing some form of lime or flurospar. The former reagent made the chemical composition of the slag more basic and therefore more suitable for eliminating such impurities as phosphorous. The latter additions were used to regulate important physical properties such as slag fluidity. One test for temperature involved observing how much the end of a steel bar melted when it was inserted into the furnace; another method relied on judging the rate at which a molten sample drawn from the bath solidified in a test spoon. As one technical manual noted, "By long practice with these methods the workmen become very expert in making these relative determinations of temperature."

After the first helper had worked the heat to the required carbon content and temperature, the melter took over and supervised the operations of tapping, teeming, and finishing the heat to specifications. However, the experience and teamwork of the furnace crew still came into play as the steel was tapped into the ladle and various elements were added to produce the exact kind and grade of steel desired. Likewise, the nozzle-setter and his helpers exercised considerable control over the final quality of the ingot as they teemed steel from the ladle into molds waiting on railway cars. The steel pourer had to prevent the stream of molten steel from striking the sides of the mold since splashes could result in uneven cooling and eventual surface defects on the ingot. Controlling the rate at which the steel filled the molds was likewise critical. Teeming from an overhead crane that carried the ladle from mold to mold called for close coordination between the crane operator

ROLLING MILLS
#4 32" universal slabbing mill. Universal mills had both horizontal and vertical rolls. The ingot was reduced in cross-section with 1 pair of grooved horizontal rolls. Three pairs of vertical rolls controlled final width (c.1905).

and the men on the pouring platform. Deoxidizing elements such as aluminum might also be thrown into the molds during teeming if a particularly uniform ingot structure was desired. The physical characteristics of ingots varied according to the amount of oxygen dissolved in the liquid steel. Aluminum suppressed the evolution of carbon monoxide and other gases that would otherwise be trapped and form blowholes as the steel cooled and solidified. Consequently, helpers on the pouring platform might be responsible for adding the proper amount of deoxidizer that would determine the eventual surface conditions and internal structure of the ingot. However, stopping gas evolution completely was only one among several methods of manipulating ingot solidification during the teeming process.

While workers' judgment played as important a role as new machinery in making open-hearth steel, the process still required muscle power as well. The charging machine and overhead crane could now carry the heavy loads of cold-charged materials or hot metal, but other ingredients still went into the furnace by hand. Charles Walker's first glimpse of an open hearth in action included ". . .men with shovels in line, marching within a yard, hurling a white gravel down red throats." As he soon discovered, this activity included making temporary repairs on the rear wall of the furnace by patching it with dolomite:

> Then all the men on the furnace and all the men on your neighbor's furnace form a dolomite line, and marching in file to the open door, fling their shovels across the flaming void to the back wall.

Making back wall required teamwork. Each man in the line watched where preceding shovel loads had landed and placed his own accordingly — not an easy task when looking into the glare through blue-smoked goggles:

> There are times when a back wall will be so cool you can hesitate a long second, as you fling your shovelful, and make sure of your aim; at others your face scorches when you first swing back, and you let the stuff off any fashion, to get out of the heat.

ROLLING MILLS
#5 48" universal plate mill. Steam engines drove table rollers while electric motors ran the screwdown mechanisms regulating distances between mill rolls. This mix of power sources was typical of early Lackawanna engineering (1904).

Open hearth crews faced other physically demanding tasks on every turn. In making front wall, another repair operation, third helpers filled a large spoon with dolomite and the second helper spread the material over the furnace banks within the spoon's 15' reach. Tapping each heat meant that the second helper had to dig out most of the material closing the tap hole; several men then broke through the remainder by inserting a long rod through the center door at the front of the furnace and then forcing it all the way across the hearth to the tap hole. As the steel poured out of the tap hole, recarburizing, deoxidizing, and alloying elements were added to the cauldron; bags of the fine anthracite occasionally used to raise the carbon content might weigh 50 lbs. Repairing the bottom of the furnace after the heat involved more manual application of basic lining materials such as magnesite or dolomite.

While there was heavy labor on both shovel gangs and furnace crews, Walker found that taking small steps up the hierarchy of job grades affected his attitude toward different types of work. After being promoted from a clean-up crew to third helper on a furnace he recalled, "For a minute I felt vastly superior to pit people. I earned 2¢ an hour more and threw down a hole the dolomite and dirt they cleaned away."

Long hours compounded fatigue from exertion and concentration. Most plants, including Lackawanna at the outset, operated their open hearths round the clock with only two turns. Continuous operation meant 10-14 hours of work seven days a week; a man changing shifts had to work 24 hours straight every two weeks — the infamous "long turn." The rhythms of work in the open hearths occasionally allowed some respites where men could try to recuperate from exhausting labor. However, as Charles Walker noted, informal rest periods had no legitimacy under prevailing industrial relations:

> You're under strain for 12 hours. Nerves and will are the company's the whole shift — whether the muscles of your hands and feet move or are still. And the existence of the long day makes possible unrelieved labor, hard and hot, the whole turn of 14 hours, if there is a need for it.

ROLLING MILLS
Hydraulic pushers replaced manual labor for charging billets into reheating furnaces (1904).

DANGER!

A first helper told young Walker one anecdote that revealed the human consequences of long hours in a physically-debilitating environment:

> Once, after a 24 hour shift, I fell asleep in the bathtub, and woke up to find the water cold.

In Lackawanna's rolling mills technological change had exerted its uneven impact on a different set of jobs and skills developed by a previous generation of steelworkers. Engineering altered many parts of the production process while leaving some jobs relatively untouched. The new rail mill offered a particularly clear example.

During the latter 19th century most American plants rolled their rails on 3-high mills where powered rollers and lifting tables handled the steel instead of men. Over the years engineers had also applied mechanical devices at other points in the overall sequence of rolling operations: machines for moving ingots or blooms at the soaking pits or reheating furnaces; hot saws to cut the rails to proper length; cambering machines to correct for warping as the rails cooled; powered hot beds where rails could be held until their temperature had dropped sufficiently for further treatment.

When the Lackawanna rail mill was being laid out, several decades of innovation had culminated in equipment that could produce more rails with fewer men. British engineers visiting U.S. plants at the turn of the century marveled at the differences between facilities on this side of the Atlantic and their own:

ROLLING MILLS
Billet reheating furnaces at the #6 16"-12"-8" bar mill. Built by Morgan Construction Co., the furnaces featured a continuous gravity end-discharge design which reduced manual handling (c.1907).

> It is in the rolling mill practice that the mechanical engineering has been seen to the best advantage, and the appliances introduced have effected the best results in doing away with manual labor. A tongs or hook is not seen near any of the rail mills visited, and the whole operation is conducted from a platform, where levers connected with the various live rollers and lifting tables are collected together. One man is in constant attendance and his duties are so heavy that two men are 'spelling' one another at short intervals.
>
> In no part of steel manufacture have inventions and improvements had such an effect upon working conditions as in the rolling mills. Twenty years ago these mills were alive with men. Today you will find large numbers of men in the guide and merchant mills, but at the blooming mills, the plate mills and the structural and rail mills you have to look sharply not to miss them entirely. These mills have become largely automatic.

Although Lackawanna's 1903 rail mill differed in detail from conventional American 3-high mills, it seemed equally awesome to untutored visitors who felt that machines had taken the place of men:

> Another feature of the mill is the almost total absence of men who are obliged to do actual manual labor. The mechanism of this plant is so wonderful that the men do not handle the rail excepting to turn it over on the floor with a set of iron tongs. The rail is shifted from one side of the floor to the other, rode on to the different sets of rollers, cut into lengths and placed in its pile, as well as being first taken from the soaking pits and loaded on the cars, all by the wonderful machinery.
>
> Hour after hour this same process goes along with scarcely a halt, and with no apparent distress the men at work in the mills perform their duties without even a thought or a heed to the fearful heat. Nearly every man employed in the rolling mill is a skilled laborer, and well he needs to be, for his work is at all times exacting . . .

As steel corporations sought technological alternatives that would increase productivity and profitability, they also reshaped daily work. However, the strategy of replacing labor with capital investments did not convert all rolling mill hands into a homogenous body of machine operators, as the preceding quotations might indicate. As at the blast furnaces and open hearths, a partially-mechanized production process meant that some jobs still required considerable manual effort.

ROLLING MILLS
#9 10" semi-continuous merchant bar mill. Men still caught steel by hand during part of the rolling operation (1911).

Within the bowels of Lackawanna's rail mill, electrically-driven rolls transported steel on the mill tables, and electrified wire rope buggies dragged the rails across the hot beds; these and other handling devices appeared on the scene as lineal descendents of friction- or steam-powered devices introduced by a preceding generation of engineers. And yet, as the newspaper story mentioned above, three men with tongs still had to turn the heavy bar over between passes on either side of the first finishing stand. Heavy lifting was even more common in straightening and drilling rails at the finishing shed where visiting journalists apparently did not venture.

On the other hand, the technological upheavals that had displaced many unskilled laborers in the mills continued to depend on some of the skills cultivated by 19th-century steelworkers. When Lackawanna Steel added rolled plates to its product line, the company found that simply installing the most up-to-date machinery did not guarantee successful operation. Consequently, Lackawanna carried out a raid on U.S. Steel's Homestead works to hire away the skilled rollers, gaugers, shearmen, and screwdown men who had already mastered current production methods. In situations where empirical knowledge and accumulated experience remained important, the difference that skill could make was obvious even to the same observers who were transfixed by visions of labor-saving machinery. One British visitor noted that the practice of making skilled rollers out of raw hands in six months was detrimental to both product quality and operating safety.

Turn-of-the-century steel technology at Lackawanna and other plants laid the foundations for a new and unique world of work, but the transformation of older methods did not create a corps of workers with homogeneous skills. In fact, the effects of uneven technological change on the overall job structure was just the opposite. Production techniques implemented at the new Lackawanna plant established a mosaic of job skills and working conditions within particular departments whose details varied considerably from process to process. However, while technology and work situations may have created self-contained islands within the plant, no worker was alone in feeling the visible hand of managerial authority on his shoulder.

Work and Authority

Lackawanna Steel's choices about the management and administration of the production processes in its new plant affected steelworkers' jobs at least as much as the new technology. Decisions concerning wages, hours, working conditions, layoffs, or promotions reflected corporate policy as well as technical requirements. The company's method of reaching and implementing those decisions was as revealing as the content of its policies. Over a 20-year period Lackawanna's attitude toward its workforce evolved from relative indifference into more active forms of corporate paternalism but, finally, regressed into outright repression when workers began organizing to bargain collectively. Shifts in company strategy mirrored changes in the steel industry as a whole as reformers began criticizing corporate labor policies that dictated long hours and hazardous conditions. Meanwhile, steelworkers from various ethnic and racial backgrounds tried banding together to secure a louder voice in decisions affecting daily work within the new job structure.

Local journalists initially reported that Lackawanna Steel employed both union and non-union workers during the construction period. "The attitude of the great corporation toward union labor is yet to be defined," though strikes by the building trades certainly tested the company's professed tolerance. However, the Board and management clarified their stance rather quickly. Mindful of the pitched battle between workers and private detectives at the Homestead plant in 1892, the newspaper reported that:

> The steel plant people favor the proposition to move Fort Porter to a point on the lake shore just above the steel plant. In case of any serious trouble a force of soldiers could be sent into the town at once.

To maintain unilateral control over the administration of production, Lackawanna Steel adopted the anti-union policies then being advocated by industry leader U.S. Steel.

Fragmented by plant technology and cultural background, workers had little success in overcoming divisions inside and outside the plant. Turnover in the physically-demanding jobs was high. The geography of the nearby community did not provide fertile ground for solidarity. Outside the plant gates, Lackawanna Steel had constructed two clumps of company housing east of the Hamburg Turnpike. The Old Village of brick row tenements, located north of Smokes Creek, was intended for workers on the lower rungs of the pay scale. The New Village consisted of residences for skilled employees, mainly detached single and double houses situated north of Ridge Road near the plant's imposing office building. Dispersed residential patterns reinforced skill and ethnic divisions among workers. Besides stratifying inhabitants, company housing could accommodate only a third of the total workforce. Steelworkers who could afford to do so dealt with the housing crunch by living in Buffalo and commuting on streetcars. Those who remained closer to the plant found alternative shelter within a maze of private boarding houses. Catering chiefly to single men and usually associated with saloons, these lodging places sprang up wherever the swampy terrain and network of railway tracks permitted. While sanitary conditions in the boarding houses were primitive, foreign workers often preferred living in quarters owned by a fellow countryman rather than the solid but sterile rows of relatively-inexpensive company housing.

Many veterans of the Scranton plant relocated to Lackawanna Steel's new works, but the initial period of construction and operation had also attracted the sizeable foreign contingent that crowded into the boarding houses. The local press perceived the new immigrants through prisms of prejudice. Polish people as a group were tainted through association with Leon Czolgosz who had recently assassinated President McKinley in Buffalo; they allegedly retained their customary veneer of radical socialism mixed with a propensity for anarchism. Italians were said to be split into cliques by traditional village loyalties and regional dialects; factional feuds carried over from the old country could easily erupt into violence. If inadequate housing or economic fluctuations made it difficult for immigrants to sink roots, their departure might not be counted as a major loss. Following the 1907 financial panic that curtailed operations at Lackawanna, a local paper observed that, "A great portion of an undesirable population has returned to Europe while the plant was down."

Blacks may have encountered an even more hostile reception. The company actively imported experienced black steelworkers from Baltimore. Fifty families arrived during the fall of 1904, and 200 more were expected. Prevailing opinions about racial characteristics colored Lackawanna's recruitment policy. "It is said, the Negroes can stand the terrific heat of the furnaces better than the white men." West Seneca town politicians occasionally manipulated racial issues, once going so far as to propose raising the steel plant's assessment in retaliation for each new batch of imported black workers. However, during the Lackawanna Steel era, there was at least one black foreman working at the plant, Joseph Moon in the rigging department. Temporary advancement for blacks in the steel industry came all too often from the companies' use of strikebreakers to keep operations going during the 1919 strike.

Lackawanna Steel's early policies regarding housing and labor recruitment reflected an overall lack of involvement with the lives of workers outside the plant gates. During 1910-1911 the company adopted a more paternalistic position in response to a growing crescendo of public citicism about labor conditions in the steel industry. Lackawanna instituted a plethora of programs then being advocated nationwide by a coalition of social reformers and corporate officials. The ideology of welfare capitalism behind these new programs urged that employers look beyond the cash nexus when attempting to minimize costly labor turnover or defuse worker militance. In recasting some of its labor policies, Lackawanna Steel established a Sociological Department similar to the bureau that later became famous at Ford. It also forged closer links with the settlement house reform movement; James Speyer, a member of the Board of Directors, happened to be active in New York City's overlapping circles of municipal reform and social service. A good deal of early 20th-century reform actively focused on housing, sanitation, and other problems associated with rapid urban growth and massive immigration — trends that

engulfed even the small city of Lackawanna when it was finally established in 1909. However, public pressure resulted in some changes that affected steelworkers on the job as well.

Long hours in the steel industry were a national scandal. Twelve-hour turns and seven-day workweeks in many departments combined with enervating conditions to create work environments that sapped the strength of even the strongest men. As one Lackawanna steelworker told visiting journalist John Fitch:

> If we had eight hours it would be a great job, with lots of excitement, one that I'd want to get back to. But the way it is now, twelve hours a day and sometimes seven days a week, I don't get any pleasure out of the job or out of life.

As of 1911 about a quarter of the Lackawanna plant's employees (1010 out of 4066) worked in departments where the process of production proceeded continuously. The men in the departments that operated seven days a week were distributed equally among the coke ovens (330), the blast furnaces (300), and open hearths (380). A seven-day week in the blast furnaces was standard throughout the industry, but Lackawanna Steel dissented from the example currently being set by U.S. Steel which did not run its open hearths around the clock. Continuous operations affected not only those who tended the ovens and furnaces but also electrical workers, millwrights, repair crews, gas-house men, and auxiliary labor gangs.

In the glare of growing public criticism, Lackawanna Steel attempted to ameliorate conditions in departments where the seven-day week had been the rule. As of 1909 a floating crew was established in the open hearths. These relief men rotated from furnace to furnace, theoretically providing each first helper and his associates with one day off every two weeks when all 12 open hearths were in operation. However, crane operators, electricians, and men working in the gas-producer plant continued on a seven-day schedule. In 1911 floating crews and spell hands were also added at the coke ovens and blast furnaces with the intent of giving each man one day's rest per week. However, off days were compulsory, assigned unilaterally by the company, and might fall on any day of the week. This arrangement sparked considerable discontent in the coke ovens so the floating gang plan there was scrapped after a trial of three weeks. Where seven-day work was eliminated, men on 12-hour turns now worked 18 consecutive hours at the conclusion of each week instead of 24 hours straight every two weeks when the day and night turns changed. Also in 1911 Lackawanna Steel reduced the working day for its 500 common laborers from 12 hours to 10. However, these modest reforms still left the minimum workweek at 60 hours for common labor and more for 12-hour workers. In addition, rates were not adjusted to compensate for the reduction in hours. Time off was scheduled at the company's discretion and workers took additional days at their own expense.

Moderate reform also focused on the issues of safety and compensation for job-related accidents during the years prior to the First World War. Throughout the 19th century, making iron and steel had been a very hazardous occupation. At Lackawanna and other modern plants companies had introduced technological changes that eliminated some of the most dangerous jobs; men no longer worked as fillers on top of the blast furnaces or stripped ingot molds on the open hearth floor. Nevertheless, the long hours combined with heavy machinery to create an environment where steelworkers still faced possible injury or death every day. For example, in February, 1904 three men working with the new gas engine system were killed; Frank Prenatt of 711 South Park Avenue in Buffalo, a 22-year-old power house worker, died in an explosion after crawling into a waste gas reservoir tank to rescue two masons who had been asphyxiated.

The prevailing interpretations of common law doctrines by conservative judges held that employers were not liable for most industrial accidents. When molten metal scalded two workers in the Bessemer shop, a local newspaper concluded that, "No blame can be attached to anybody for the splashing of metal in a steel works, it is said. It is one of the risks which the steel maker takes." However, suits for damages on grounds of company negligence had been filed as early as 1902, and at least one worker's claim was upheld in court during 1907.

Making ingot molds in collapsible-core barrels at foundry. This procedure saved the expense of casting ingot molds in rigid-core barrels. In steelmaking departments molten steel would later be teemed into molds to form ingots.

Lackawanna Steel attempted to rationalize its procedure for dealing with injured workers' claims during the same period that it reduced hours in some departments. Originally, the corporation's police department, modeled on the organization at Illinois Steel's Joliet works, had also been responsible for investigating accidents. In 1906 state officials had looked into rumors that Lackawanna was not reporting some accidents to the proper authorities. By 1911 the company had formed a Safety Department whose inspectors dealt with the prevention of accidents as well as the consequences. They had plenty of opportunities to keep busy; during 1910-1911 when the average number of plant employees varied between 4957 and 4066 there were 28 reported fatalities, 252 accidents requiring hospitalization, and 1612 accidents where workers had to be sent home. The Safety Department claimed that its new regulations had reduced the number of accidents when compared with the situation five years earlier.

However, as of 1912, Lackawanna Steel still had not formulated fixed standards of compensation for employees injured or killed on the job. Settlements varied from case to case. The corporation did perceive a need for setting such standards to ". . . reduce the misunderstanding of our employees and also facilitate the prompt settlement of all cases." Lackawanna Steel officials were also active in reforming state legislation affecting workers' compensation. Such laws were part of a broader program put forward by some businessmen to persuade labor that its best interest lay in cooperating with employers, even though employers would still define the issues where cooperation was possible.

Traveling coal hopper in #2 boiler house. This machine served 48 coal-fired Cahall vertical water tube boilers. 32 other boilers behind the hopper burned blast furnace gas instead of coal (c.1904).

Despite some marginal improvements due to reforms advocated by other interest groups, Lackawanna's early steelworkers had little direct voice in the decisions that affected their jobs and lives. Internal divisions hampered collective opposition to wage reductions and other corporate policies that could not be camouflaged as paternalism. When the company cut the salaries of office workers by 10% in March, 1909 and lowered the basic common labor rate by a half a penny to 12¢ an hour, it was noted that "Wage agreements complicate the situation very little, either here or elsewhere." In discussing the significance of the brass time checks issued by the company, a local newspaper story commented indirectly on the anonymity of individual workers in the new world of complex technology and corporate power. "After assuming this number the laborer loses his name and thereafter is known only by the figures on his coat."

A character in Thomas Bell's classic novel, *Out of This Furnace*, articulated his perception of the psychological gulf separating steelworkers from the authority figures above them:

> . . . the General Superintendent wasn't an ordinary man set a little above others much like himself. He was of a different species altogether, partaking of the company's inhuman incomprehensibility and like the company going his own way, doing this, doing that, hiring, firing, shutting down, cutting wages, raising them, and giving no reasons, maintaining throughout a cold impassibility before which a man felt himself divested of humanity . . .

During and after the First World War steelworkers at Lackawanna and other plants began organizing to strike a more equitable balance of power in industrial relations. Wartime conditions favored this movement at first. The conflict in Europe dammed the steady stream of immigration, creating a labor shortage as production boomed. As the pool of available workers dried up, in 1916 Lackawanna Steel attempted to roll back its recent policy of eliminating seven-day labor, but ran afoul of a New York law prohibiting the old practice; the company claimed that such legislation gave competitors in other states an unfair advantage. However, neither lengthening hours nor importing more black workers from the boll weevil-infested fields of the South could ease the labor crunch. Steel companies struggling to maintain or expand their work forces began raising wages and hinting at other concessions.

Roll shop. "Rollturning is an art that few may learn. It is usually transmitted from father to son." *Buffalo Express*, 2 February 1905.

A factor even more important in establishing a climate where change seemed possible was the wartime activity of the federal government. Agencies such as the National War Labor Board and the War Labor Policies Board came out in favor of collective bargaining and against the steel industry's 12-hour day. In perhaps the most famous case, the War Labor Board ordered Bethlehem Steel to establish collective bargaining with a system of shop committees in 1918; other policies also threatened the bastion of the open shop, keystone of the existing authority structure in the steel industry. Building on momentum generated by these wartime changes, the uphill battle to make unions a legitimate part of the steel industry got underway during the summer of 1918 and reached Lackawanna by the end of the year.

The strategy for organizing the steel industry rested on a joint effort by craft unions who were members of the American Federation of Labor. The local committee in the Buffalo-Lackawanna area included delegates from 19 unions: Amalgamated Association of Iron, Steel and Tin Workers; Blacksmiths and Helpers; Boilermakers and Iron Shipbuilders; Bricklayers, Plasterers and Masons; Bridge and Structural Iron Workers; Coopers; Electrical Workers; Longshoremen; Machinists; Molders; Patternmakers; Plumbers and Steamfitters; Railway Carmen; Seamen; Sheet Metal Workers; Stationary Firemen; Steam Engineers; Steam Shovel and Dredgemen; Switchmen. Several of these craft organizations could have paralyzed steel plant operations by organizing key occupations under their jurisdiction. For example, the railway unions and the electrical workers, whose territory included cranemen, aimed at recruiting men responsible for moving the materials that kept an integrated plant going. However, the decentralized structure, internal dissention among some unions, and an overall deficiency of resources handicapped the organizing effort.

Organizers often encountered hostility from public officials and police in company towns, particularly in the area surrounding Pittsburgh. While the campaign in Buffalo got underway without much opposition, it immediately ran into obstacles in Lackawanna. The city's mayor and police commissioner refused to issue a permit for a meeting scheduled at Croatian Hall, 420 Simon Avenue, on 19 December 1918. Petitions lodged with the city council protested this denial of the constitutional right of lawful assembly as contrary to the aims for which the Great War had supposedly been fought. However, it was not until early August of 1919 that the local iron and steel organizing committee was able to hold a meeting in Lackawanna. During March the city council had considered a municipal ordinance stipulating that all remarks at mass meetings must be spoken in English unless otherwise authorized by the mayor. Nevertheless, the first meeting held at the A.C. Theater, Electric and Kirby Avenues, on 2 August did include a Polish speaker, Thomas Rokois. A local organizing office was also opened at 412 Ingham Avenue. While the theater was not the most convenient site for meetings (one report referred to it as being located "out in the goose pastures"), the partial relaxation of local political pressure permitted the steelworkers' campaign to reach more native and immigrant employees of the Lackawanna plant.

In mid-August a Lackawanna meeting approved the strike issues formulated by the national organizing committee: recognition of collective bargaining; reinstatement with back pay of any employees discharged for organizing; the 8-hour day; abolition of 24-hour shifts; one day's rest out of seven; wage increases; standardization of wage scales for all crafts and classifications; double time for overtime, Sundays, and holidays; a check-off procedure for deducting union dues directly from members' paychecks; recognition of seniority as a basis for determining layoffs and promotions; abolition of company unions and employment physicals. Enacting that agenda would take another three decades. Issues which may seem moderate and reasonable today were viewed by leaders of the steel industry as radical if not subversive.

On 22 September 1919 rank-and-file pressure escalated the year-long national organzing campaign into a national strike. Judge Elbert H. Gary of U.S. Steel, the industry's leading spokesman, precipitated the conflict by refusing to engage in any negotiations. There was violence in Lackawanna at the outset. On 24 September shots fired by private plant guards killed Kasimir Mazurek and injured three others; the driver of a laundry wagon was clubbed while making his rounds. Lackawanna's Mayor Toomey summoned the state police, a maneuver that backfired and cost him his job in the November election. Some of the police behaved badly. Periodic scuffles continued to erupt, particularly on the streetcar lines transporting strikebreakers, including some of the 2000 blacks hired by the company. But the conflict eventually settled down to a war of attrition, especially after the Industrial Conference convened in October by President Woodrow Wilson failed to reach any compromise solution. As the prospect of federal intervention evaporated, the national and local strike committees began undertaking relief efforts to sustain their members over the long haul.

The strike in Lackawanna remained effective longer than in most districts. As winter closed in, production in the steel plant slowly resumed, but as late as mid-December thousands of strikers still packed Dom Polski Hall on Ridge Road where speakers exhorted men in English, Serbian, Croatian, and Polish. On 8 January 1920 the national committee decided that further opposition would be futile. However, as many as 3500 Lackawanna workers continued an independent strike for another two weeks.

While the industry's deeper pockets eventually prevailed in this major conflict, Lackawanna Steel itself suffered considerable financial loss. When the cycle of protest started up again in the mid-1930s, the story in Lackawanna would involve a different corporation as well as altered economic circumstances, organizing tactics, and styles of government intervention.

BETHLEHEM STEEL 1922-1970

4. THE CORPORATION

"Bear in mind that when I took Bethlehem it was a bankrupt concern, its plants were run down and obsolete, its buildings ramshackle . . . When I helped form the United States Steel Corporation, I took all the best. When I organized Bethlehem, I took all the leftovers and made them into the best."

This immodest pronouncement by Bethlehem's founder, Charles Schwab, gives us insight into the company and the man who ran it; Schwab established policies and practices which survived long after his death in 1939 and which shaped Lackawanna's heyday as a premium facility within a rapidly-growing corporation.

Unlike Lackawanna's original owners, Schwab was a steelman. One of business history's rare genuine rags-to-riches paragons, he began his career as a common laborer at Andrew Carnegie's mills. He moved rapidly through the ranks, finally becoming president of the new U.S. Steel Company. Although he once broke with both Carnegie and U.S. Steel (they reconciled by 1906), his experiences at the nation's hegemonic steel combine shaped his determination to make Bethlehem a worthy rival to his former employer.

In 1904 Bethlehem was no challenge at all; Schwab called his new company "a wreck financially and physically." Capitalized at $30 million, it was burdened by numerous unprofitable operations which Schwab sold. The new reorganized corporation rested on three foundations: the steel works, raw materials supplies, and shipyards, all of which Schwab improved. He announced that he was going to surpass U.S. Steel to make Bethlehem "the greatest plant in the world." While that feat was never accomplished, by 1910 the company's fixed assets had more than doubled from $18 million to $39.9 million, then doubled again by 1915. The company became the nation's second-largest steel producer by 1916, and by 1921, the year before it acquired Lackawanna and several other firms, Bethlehem's ingot capacity was already 6% of the nation's total.

Bethlehem's corporate strategies were similar to those of Lackawanna Steel — expansion, full vertical integration, modest diversification. The 1922 acquisition of first Lackawanna, then Midvale Steel & Ordnance, were evidence of Schwab's and the company's determination to increase its horizontal bases of control over steel production. Schwab's personal philosophy had likely been shaped by his years at U.S. Steel, and he openly advocated the consolidation of all American steel firms into one giant combine. "The greatest economy that can come to this country is doing business or banking on the largest scale possible . . . I think the idea of the United States Steel Corporation, as it originated with me, was an economic one."

The goals of the merger with Lackawanna were to acquire operations Bethlehem did not already have: merchant steel bars; increased rail production for new markets; improved production of specialty products. The new combined ingot capacity would be 5 million tons, or 10% of national output. Despite Bethlehem's greater size and the neglect of Lackawanna's basic plant, the latter firm outproduced Bethlehem in several key areas including sheet bars, standard and light-tie rails, standard structural shapes, and several dimensions of plates. In addition, Lackawanna's superior control over domestic ore and fuel sources improved on Bethlehem's raw materials supplies which were then primarily from Cuba.

The Lackawanna and Midvale merger with Bethlehem involved 46 separate companies under three groups. Bethlehem Group had 11 companies; Lackawanna had 13,

Bethlehem Chairman Eugene G. Grace and top company officials meet in his office at corporate headquarters in Bethlehem, PA (c.1950).

and Midvale owned 22. Each group was an integrated concern controlling all stages of production. This merger was only the first in a series of moves taken by Bethlehem to increase its overall corporate strength. Between 1930 and 1936 alone, in the depths of the Depression, the company attempted no fewer than 11 separate mergers, 8 of which succeeded. Earlier, before 1930, the company had acquired 11 separate steel plants of various sizes, with 5 in Pennsylvania, one — Lackawanna — in New York, one each in Delaware and Maryland, two in California, and one in Seattle, Washington. Part of the Depression-era acquisition program included purchase of steel fabricating operations such as McClintic-Marshall Construction, Kalman Steel, and 3 other firms. In 1932 Bethlehem expanded its steel finishing operations by acquiring Seneca Iron and Steel Company near Lackawanna, then in 1937 purchased the bankrupt Williamsport Wire Rope Company in Pennsylvania. During and after World War II, Bethlehem added Pacific Coast Forge, a steel barrel manufactory, and two firms producing oil-drilling equipment. Thus the company worked to expand its basic steel production, finishing, and fabricating operations throughout the Depression, wartime, and the post-war aftermath.

The merger with Lackawanna was also important to Bethlehem for its contribution to increased backward integration. Following the merger, Bethlehem began expanding domestic reserves on its own, adding ore properties in the Michigan Gogebic Range as a wholly-owned subsidiary. Other partially-owned properties were located in the Marquette, Menominee, Cuyuna, Vermillion, and Mesabi ranges. Bethlehem continued to expand properties abroad in Chile, Mexico, Venezuela, and Cuba, all under government leases. In addition to ore, Bethlehem developed domestic coal reserves in Pennsylvania and New Jersey. In the mid-1950s Bethlehem entered a major international ore venture, the Erie Mining Company. Bethlehem held the majority 45% share with the remainder divided among Interlake Iron Company, Steel Company of Canada, and an old Lackawanna Steel ally, Pickands-Mather of Cleveland.

Forward integration consisted chiefly of transport operations as had Lackawanna's; by 1950 Bethlehem owned seven short-line railroads, all servicing steel or shipbuilding operations. Under three large sub-divisions the company owned eleven ocean-going ore carriers and two general cargo vessels plus ten lake vessels and two self-unloaders for product as well as raw materials transport. To facilitate their transport activities they added a dock company and opened an export subsidiary to connect steel production with shipping operations and to handle overseas sales.

The efforts to diversify remained fairly peripheral during the middle period. Before 1970 the company conducted virtually no overseas direct manufacture. Domestically, their non-steel-related holdings were few in number; they owned, at various times, a county utility company in Lebanon, Pennsylvania, a sewer company in Maryland, and the Farmers & Merchants Bank in West Virginia. However, their non-steel ventures were by no means true diversification, since even these modest operations were designed to enhance the production of steel.

Corporate Practices

Bethlehem's quest for market dominance resulted in continuation of corporate practices we have seen at Lackawanna. A member of the American Iron & Steel Institute and a fixture at the Judge Gary dinners, Bethlehem and its top managers worked cooperatively rather than competitively with other industry giants. Throughout the whole of its middle period the company was all-too-frequently charged with price fixing, with establishing collusive production and marketing agreements, and with contravening anti-trust regulations.

In January, 1930 reports in the *New York Times* stated that "certain American steel concerns, notably the Bethlehem Steel Corporation" were considering joining the European steel cartel. The impetus for this move, a potential violation of U.S. law, came from fears of foreign competition; this strategy would extend the existing practice of U.S. participation in individual product-line cartels. By 1936 Bethlehem's involvement became more open as they and U.S. Steel were invited by British cartel chief, the Earl of Dudley, to join the powerful European association. Bethlehem President Eugene G. Grace openly advocated participation. Foreign imports, particularly from

Germany, were already a problem in the U.S. markets; Bethlehem had lost a major contract from the Public Works Administration to build the Tri-Borough bridge. The project went instead to the Kloeckner Steel Company of Germany. Other evidence of foreign competition plagued the company. In 1938 they had lost an important iron-ore leasehold in Mexico which was given instead to Japan in exchange for that country's modernization of the port of Mazatlan. It was later confirmed that in 1938 Bethlehem accepted the Earl of Dudley's offer and joined the cartel, although participation was clearly ended with the war.

Following the massive production boom which accompanied World War II, Bethlehem once again began manufacturing steel for the domestic market. Post-war growth was enormous; the demand for new national construction was exceeded only by that for overseas reconstruction. The company did expand its export subsidiary to fill new market orders, but their overseas operations continued to be mainly in raw materials rather than finished goods or direct production.

In 1950 the industry as a whole was warned that it was losing important markets in South America as energetic, modern firms from Germany, England, and Japan competed there more successfully, offering not only somewhat lower prices but much better terms of credit. No U.S. steel company met this challenge, however, and the American steel industry continued to rely on domestic rather than foreign expansion. This was unquestionably Bethlehem's peak period of production growth; doing what they did best, operating as "a steel company, pure and simple", their sales volume grew steadily, the company made profits, and plant capacity expanded at a rapid pace.

Debt

The high price of expansion dogged Bethlehem throughout its middle period. The company initially avoided acquiring large debts to finance expansion, but it inevitably became responsible for the outstanding debts of its subsidiaries. Both Lackawanna and Midvale came with a hefty package of debts for which Bethlehem had to pay; Lackawanna had reduced some of its obligation but still owed well over $21 million. While the actual cash outlay for the merger was negligible — it was accomplished through a stock swap — the unpaid loans were a real burden. Midvale added another $50 million to the pool of indebtedness in 1923.

As Lackawanna Steel had done initially, Bethlehem pursued several logical steps to refinance these loans. First they consolidated the multiple debts into more manageable packages with lower overall interest, then they used the interest savings plus retained earnings to retire complete bond series and even whole debt packages. This freed still other funds for plant modernization and further development. Between 1925 and 1935 Bethlehem cut its funded debt by more than 50%, then in 1936-1940 they issued new consolidated mortgage bonds to retire other loans. These late 1930s bonds amounted to $500 million, a considerable debt in its own right, but by 1955 the entire debt had been reduced to under $150 million.

Beginning in the mid-1950s, however, the drive for expansion began to reinflate the debt package. With modernization and plant expansion, particularly the 1960s addition of the new Burns Harbor plant in Indiana, debts piled up at record levels. Between 1950 and 1970 the face value of Bethlehem's loans escalated 339% from just under $500 million to over $1.7 billion, and debt reorganization became a thing of the past.

The large sums lent to Bethlehem over the company's middle period, 1923-1970, indicate the firm's centrality within corporate America and the degree of faith large banks had in the steel company's solvency. Bethlehem retained, then strengthened, the banking alliances begun by Lackawanna Steel. Bethlehem's largest creditors were among some of the biggest financial institutions in the United States. Morgan Guaranty Trust and Chemical Bank, both within the top 10 banks in the nation, each lent Bethlehem $500 million; other large lenders included Chase Manhattan, First National City Bank (Citicorp), and Philadelphia's Girard Trust plus several smaller banks. By 1970 the outstanding $1 billion-plus loans came from only four banks: Morgan, Citicorp,

Chemical, and Chase, each of which clearly believed in the steel company's ability to repay those loans.

Ownership and Control

Bethlehem's policy-making and management structure passed through three distinct phases. The first was the "Schwab Era", already in place when the company absorbed Lackawanna. The second, following Schwab's death in 1939, was a period of insular management and little outside influence. The third period which began in 1967 and lasted the rest of Lackawanna's life was characterized by stable management coupled with renewed outside influence which served to shape corporate policies in ways different from earlier practices.

Under Charles Schwab, Bethlehem's management combined old-fashioned entrepreneurship with corporate management under a Board of Directors. Schwab indisputably exerted significant direct, personal control over company policies in concert with a friendly Board. He developed a more diversified management structure than had Lackawanna, but it was nonetheless hierarchical, with authority vested in a handful of top managers and the Board. The stability of this structure was evidenced by the comparatively low turnover in personnel who carried on the policies and practices Schwab had established.

The early Board of Directors was similar to Lackawanna's; there were both inside directors who worked for Bethlehem and outside directors whose primary affiliations lay elsewhere. Schwab, Grace, and two Vice Presidents, Archibald Johnson and Henry S. Snyder, were the first inside directors on the newly-reconstituted Board following the 1922 merger with Lackawanna. The outside directors represented various corporate and financial interests, but they were a more diverse lot than had guided Lackawanna. There were still bankers but fewer railroad men; more directors from potential Bethlehem customers and suppliers but fewer independent entrepreneurs.

The ties to the Morgan interests were still in force on this 1923 Board. Schwab was an old friend of J.P. Morgan; it was Schwab who convinced Morgan to buy Carnegie Steel and form U.S. Steel which Schwab then ran. Harold Stanley and Grayson M.-P. Murphy were both directors of Guaranty Trust, the Morgan ally, while Murphy was also a director of New York Trust. Other interests prevailed, too. Former Lackawanna Chairman Moses Taylor continued on the Bethlehem Board, bringing his National City Bank affiliation as well. However, the Guaranty connection was the most important since the bank provided the $500 million loan Bethlehem needed to modernize and expand Lackawanna and Midvale as well as to begin retiring their debts. Schwab, conversely, sat on the Rockefeller-linked Chase National board.

The primary non-financial ties between Bethlehem and other companies in 1923 involved firms which were potential suppliers of basic goods needed in steel production. Among these were Sinclair Consolidated Oil, Ludlow Valve (Troy, New York), U.S. Industrial Alcohol Co., Chicago Pneumatic Tool Company, and the major rubber companies, B.F. Goodrich and Goodyear.

By 1930 the steel company's ties had expanded. The connection to Guaranty Trust strengthened as Guaranty President William C. Potter replaced Stanley, while G.M.-P. Murphy continued on both boards. In addition, Percy A. Rockefeller, cousin to John D. II, joined the Board in his capacity as director of the James Stillman/Rockefeller-dominated National City Bank. Percy Rockefeller replaced Taylor, who had retired. In 1940 National City funded a new $35 million loan for Bethlehem after acquiring Lackawanna's old affiliate, Farmers Loan and Trust.

Rockefeller was also a director of the private investment bank, W.A. Harriman & Company, later to merge with Lackawanna affiliate, Brown Brothers, and was a director of the Rockefeller-owned Mesabi Iron Company, a critical ore supplier. Other raw goods and materials suppliers on the Bethlehem Board were from Rosiclare Lead & Flurospar Mining, Pierce Oil and Pierce Petroleum companies, and United Metals Selling Company. Potential customers included International Motor Truck Company, the Baldwin Locomotive Works, New York Railways Company, New York Transportation Company, and the Lackawanna ally, the Atchison, Topeka & Santa Fe. All of the railways were potential customers of Bethlehem's rails and related products.

Complex ties between the Morgan and Rockefeller cliques interlocked them with Bethlehem as they had with Lackawanna. Percy Rockefeller was not only part of his family's circle of influence but was also a director of Anaconda Copper, a close Morgan affiliate. Baldwin Locomotive was part of the Morgan circle as well.

Following Schwab's death and the rise of President Eugene G. Grace to the Chairmanship, Bethlehem radically reconstructed its Board and therefore the bases of policy formation. The Boards for 1940, 1950, and 1960 included only one outside director (1940). For 27 years Bethlehem was a company of and for steelmen. It was also the period of the company's greatest growth and profitability, so that neither the company nor, apparently, the banking community saw a need for outside direction, even though the steel company remained tied to major banks through its loans.

However, in 1967, when the corporate community in general and the manufacturing sector in particular began to feel the first effects of post-war slowdown, Bethlehem once again placed outside directors on the Board. It is hardly surprising that the reconstituted Board included representatives from Guaranty Trust (now "Morgan Guaranty" after a formal 1959 merger). Chemical Bank, a Bethlehem creditor since 1953, and Continental Illinois also gained representatives on the Board. Continental had no loan arrangements with Bethlehem but was the leading mid-west bank and was naturally interested in their new neighbor at Burns Harbor, Indiana.

Once again Bethlehem was close to the center of corporate power. Through both Morgan and Chemical banks they indirectly interlocked with several other steel companies: U.S. Steel, Republic, National, and Lykes-Youngstown. Bethlehem director and executive Stewart S. Cort sat on the Continental Board of Directors along with Chicago-area rival, Inland Steel. Other major corporations on the late 1960s Bethlehem Boards were Chrysler, Cummins Engine, AT&T, Equitable Life Assurance, Black & Decker, Nabisco, and McDonnell-Douglas. Clearly, after years of both formal and informal ties to the centers of finance and industry, Bethlehem once again realized Schwab's goal of making the once down-at-the-heels steel company into a major corporate power.

Profitability

The extension of loans and the interlocking ties among major corporations reflect Bethlehem's strong performance record during its middle period. Even the Depression barely slowed the company; Schwab firmly believed that there was no better time to expand than in periods of general decline, and his strategy proved correct.

Nevertheless, while the company almost always earned profits, the rates of return were highly volatile. As with any corporation, Bethlehem was affected by the domestic and global events of Depression, war, and post-war recovery as well as factors such as tax reform, environmental regulation, labor demands, and the like. Profits were also influenced by internal corporate decisions such as pricing policies. It is here that we see the ultimate effects of Bethlehem's cooperative rather than competitive relations with other major steel companies, effects which partially shaped the rates of return and overall profitability.

Bethlehem, as Lackawanna before, carried on the practice advocated by Schwab and others of administering prices rather than competing in terms of supply and demand. Between 1930 and 1970, prices per ton on all steel and on fabricated products had a tendency to rise, even in periods of low demand. In April, 1962, when industry-wide prices did temporarily roll back at the insistence of the Kennedy administration, Bethlehem Chairman A.B. Homer told the *Wall Street Journal* that Bethlehem had done so only because two other companies had "refused to follow the increase". This was a clear enunciation of the cooperative, if not collusive, nature of steel price setting.

Even during the Depression prices held relatively steady, although at one point Bethlehem's capacity was running at only 16% and Lackawanna's at only 2%. In 1931 prices rose, then rose again as rail prices went up in 1932. Across-the-board increases were made in 1936 and 1937, held steady in 1938 only to be reduced slightly in the last half of that year, then rose substantially in 1941.

Post-war prices rose more consistently. Between 1946 and 1958 prices rose an average of $5.74 per ton each year, with increases of $7.35 per ton and $8.50 per ton in 1955 and 1956 respectively. Between 1962 and 1970, prices rose 20 times and were lowered only 4 times. In 1969 reinforcing bar prices were raised twice for a total increase of 15.3% while that same year prices increased for other bar products, semi-finished steel, wire rod, structural shapes, and plates. These hikes were then followed by an across-the-board price rise. The previous year the company maintained an unyielding position on progressive increases, declaring publicly that they would not cut prices to stem the incursions of low-cost imports.

Earnings throughout the 40 years, 1930-1970, tended to increase, largely pulled by prices rather than pushed by cost-cutting. The company made substantial profits 1929-1930, lost money 1931-1933, then regained a straight run of profitable years 1934-1940. Of these last years the best were 1937 and 1940, even before the full effects of war production took hold.

During World War II the company showed record income with gross sales reaching the billion-dollar mark for the first time: $1.9 billion in 1943 and $1.7 billion in 1944. Sales crossed the $2 billion mark in 1953 and by 1958 became the company norm. This sales growth seems extraordinary, but it was typical of the post-war boom which lasted more than 25 years and which set historic records in sales and profits.

However, basic manufacturing did not fare as well in terms of profit rates as newer, more technologically-advanced and capital-intensive industries such as nuclear energy, computers, and electronics. Basic industries rarely achieved rates of return higher than 5-10% even when overall corporate rates averaged 16%, and some companies consistently made 20-30% per year. Steel companies certainly could boast steady but never really spectacular profit rates.

Bethlehem's best rates of profit were in its early years after building up its corporate holdings. By the Depression, rates began to vary widely. In 1930 the company's actual earnings on sales equalled 9.8%, followed by three years of losses. The 1934 recovery began with a profit rate of only .3% but later returns were much improved, soaring to 7.6% in 1937 with a profit of $31.8 million on $417.5 million in sales.

Ironically, despite the record volume and dollar value of wartime sales, Bethlehem's worst rates of return, without actual losses, were in 1943 and 1944. In 1943, on sales of nearly $2 billion they earned only $32.1 million or 1.7%. In 1944 they sold $1.7 billion but earned only $36.1 million or 2.1%.

Post-war sales remained high and dollar values of course increased, but rates of return always remained below the coveted double-digit figures.

Post-War Earnings ($ in millions)

Year	Gross Sales	Net Proft	Rate (%)	Year	Gross Sales	Net Profit	Rate (%)
1948	1,312.6	90.3	6.9%	1958	2,005.9	137.7	6.9%
1949	1,266.8	99.3	7.8%	1959	2,055.7	117.2	5.7%
1953	2,082.0	133.9	6.4%	1968	2,089.0	160.5	7.7%
1954	1,656.8	132.8	8.0%	1969	2,095.0	156.5	7.5%

Costs

While Lackawanna pushed profits by cutting costs, Bethlehem did not follow the same practice. In part their costs became more fixed; in part the company simply found it easier to raise prices. Primary corporate expenses are the cost of labor, the costs of goods purchased, and the costs of capital — or, as we more commonly call it — interest rates.

Labor costs are always the most volatile and often controversial parts of corporate expenses. The establishment of wage rates in the pre-union era was very much a matter of company volition; affected by relatively few external forces except labor shortages, the company was free to set rates in terms of their own profit demands. However, when unions won the right to bargain collectively for workers, the terms and conditions of wage setting changed. Wages after 1942 were established by agreement between workers and the company and were set for fixed periods by unbreakable contracts.

Bethlehem's antipathy to unions and to sharing power with workers was legendary. Schwab had declared early in the company's history that any man could come to him and bargain individually based on the worker's merit, but he refused to bargain collectively. So strong were company sentiments against unionization that in the 1920s Bethlehem was sued by several contractors for refusing to sell steel to any company with union workers.

Theoretically, the union era should have increased Bethlehem's labor costs as workers pressed for wage gains and resisted cuts. However, that is not precisely what occurred. Between 1930 and 1970, Bethlehem's wage costs, calculated as a percentage of gross sales, actually fell, even as the dollar rate increased. By 1960 the percentage of sales dollars going to wages was actually 6.6% less than it had been in 1930.

During the first 8 years of the Depression, 1930-1938, wage costs remained above 40% of sales, despite the fact that the company cut wages and laid off record numbers of employees. In 1930, wages were 44.6% of gross sales, falling steadily to a low of 40.6% in 1936, then rebounding to 44.5% in 1937. In 1939 and 1940, wage costs fell to lows of first 38.3% then 35.2%, but by the mid-war years of 1943 and 1944 they rebounded to record highs of 47.1% and 48.5% respectively, despite wage regulations by various wartime labor and production boards at the federal level.

Post-war recovery, complete with aggressive union demands for wage increases, never saw wage costs at that level again. In the 1950s wages remained in the 30th percentile, rising in 1958 to a decade high of 39.8% before falling 2% the next year and even more in 1960. For the next decade wages hovered within the upper 30th percentile but would not reach 40% of sales until the final decade of operation where they stayed until 1983. Thus even in Bethlehem's boom period when both prosperity and union strength boosted per capita earnings to all-time highs, wage costs to the company were actually declining in relation to sales figures and never again equalled the rates prevalent in earlier years of the company's history.

Costs for raw materials and other supplies tended to remain static. A corporation's chief goal in building full integration is to reduce costs of the goods needed to create production and/or to sell its products. The more vertically-integrated the firm, the less it has to pay to outside suppliers.

Bethlehem's level of integration was achieved relatively early after it absorbed Lackawanna. Between 1922 and 1970, the company spent nearly $1.2 billion (73%) in upgrading and modernizing its basic steel capacity but only $442 million (27%) in acquiring integrated facilities. This ratio is healthy in terms of the overall national economy since mergers, especially of integrated steel firms, threatened the competitive balance of the industry already challenged by cooperative pricing agreements. For this reason the Justice Department refused several times to allow Bethlehem's merger with Youngstown Sheet & Tube. However, Bethlehem's decision not to pursue increased diversification or even integration meant that the proportion of expenditures for goods needed in production remained static at about 40% of the sales dollar. As Bethlehem's own prices rose, so did those of their suppliers so that the steel company had very little margin for profit expansion in this direction.

The cost of capital for expansion and modernization was the third factor affecting Bethlehem's profitability. Despite a good track record of debt reduction before World War II, Bethlehem's interest costs began to increase markedly after 1961. Between 1955 and 1961 the company had reduced interest charges by more than half ($8.3 million down to $4.0 million), but expansion plans, especially the addition of the Burns Harbor facility, reversed that trend by 1963. The proportion of interest as a percent of total assets (not just annual sales) more than doubled in six years, 1963-1969 from 7.2% to 17.%.

Finally, a fourth factor contributed to the drain on earnings. The company's expansion plans were not funded by debt alone. Periodically the company issued new stock as a source of capital without interest charges. Unlike bondholders, stock owners do not absolutely have to be paid in times of low profits, and there were years when stockholders received no dividends at all. However, increased stock issues meant that

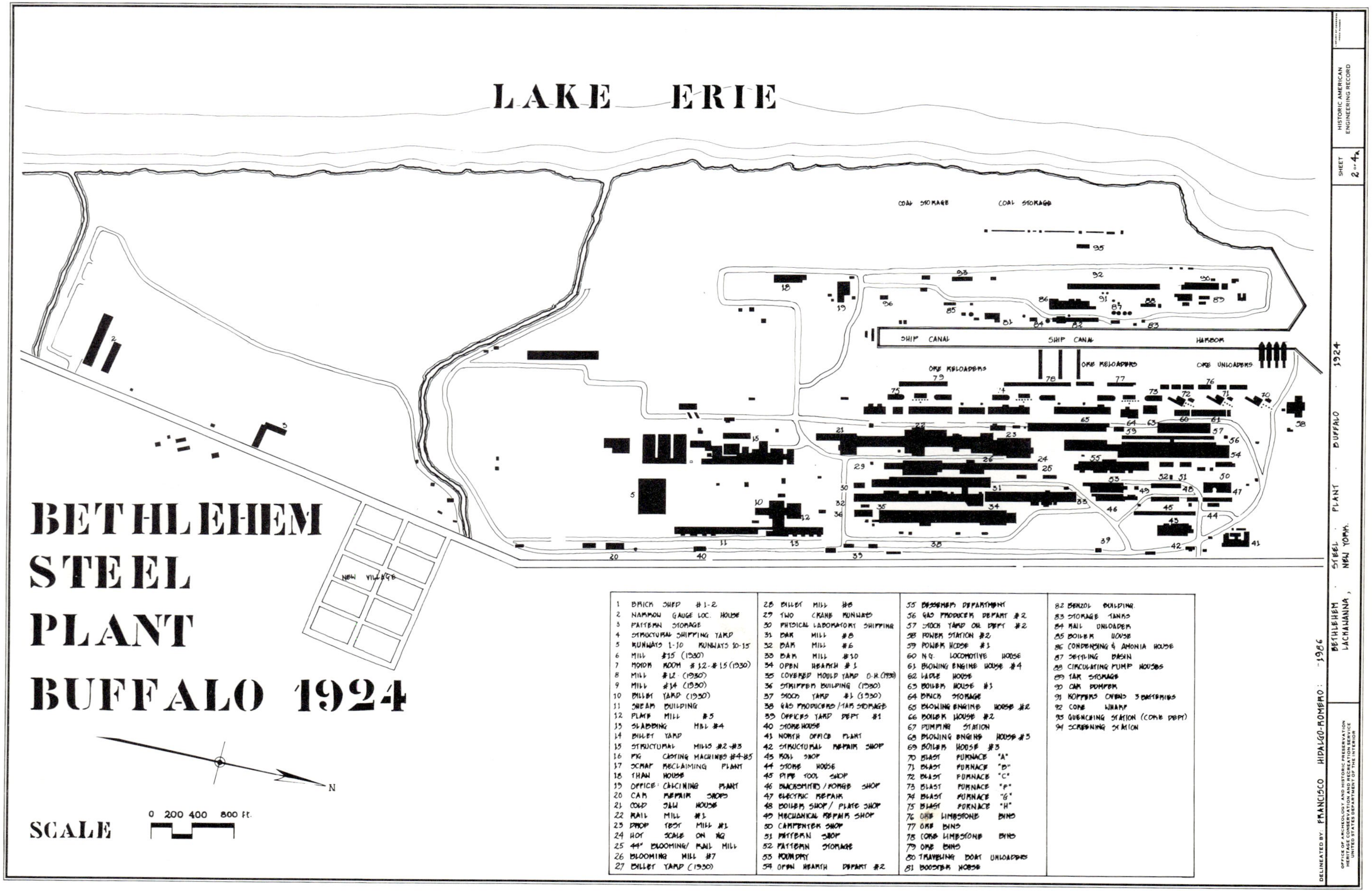
LAKE ERIE
COAL STORAGE
COAL STORAGE
SHIP CANAL
SHIP CANAL
HARBOR
ORE RELOADERS
ORE RELOADERS
ORE UNLOADERS
NEW VILLAGE
BETHLEHEM
STEEL
PLANT
BUFFALO 1924
N
SCALE
0 200 400 800 Ft.
1 BRICK SHED #1-2
2 NARROW GAUGE LOC. HOUSE
3 PATTERN STORAGE
4 STRUCTURAL SHIPPING YARD
5 RUNWAYS 1-10 RUNWAYS 10-15
6 MILL #15 (1930)
7 MOTOR ROOM #12-#15 (1930)
8 MILL #12 (1930)
9 MILL #14 (1930)
10 BILLET YARD (1930)
11 SHEAR BUILDING
12 PLATE MILL #5
13 SLABBING MILL #4
14 BILLET YARD
15 STRUCTURAL MILLS #2-#3
16 PIG CASTING MACHINES #4-#5
17 SCRAP RECLAIMING PLANT
18 TRAN HOUSE
19 OFFICE: CALCINING PLANT
20 CAR REPAIR SHOPS
21 COLD SAW HOUSE
22 RAIL MILL #1
23 DROP TEST MILL #1
24 HOT SCALE ON RG
25 44" BLOOMING/ RAIL MILL
26 BLOOMING MILL #7
27 BILLET YARD (1930)
28 BILLET MILL #8
29 TWO CRANE RUNWAYS
30 PHYSICAL LABORATORY SHIPPING
31 BAR MILL #8
32 BAR MILL #6
33 BAR MILL #10
34 OPEN HEARTH #1
35 COVERED MOULD YARD O.H. (1930)
36 STRIPPER BUILDING (1930)
37 STOCK YARD #1 (1930)
38 GAS PRODUCERS / TAR STORAGE
39 OFFICES YARD DEPT #1
40 STORE HOUSE
41 NORTH OFFICE PLANT
42 STRUCTURAL REPAIR SHOP
43 ROLL SHOP
44 STORE HOUSE
45 PIPE TOOL SHOP
46 BLACKSMITHS / FORGE SHOP
47 ELECTRIC REPAIR
48 BOILER SHOP / PLATE SHOP
49 MECHANICAL REPAIR SHOP
50 CARPENTER SHOP
51 PATTERN SHOP
52 PATTERN STORAGE
53 FOUNDRY
54 OPEN HEARTH DEPART #2
55 BESSEMER DEPARTMENT
56 GAS PRODUCER DEPART #2
57 STOCK YARD OH DEPT #2
58 POWER STATION #2
59 POWER HOUSE #1
60 N.G. LOCOMOTIVE HOUSE
61 BLOWING ENGINE HOUSE #4
62 LADLE HOUSE
63 BOILER HOUSE #1
64 BRICK STORAGE
65 BLOWING ENGINE HOUSE #2
66 BOILER HOUSE #2
67 PUMPING STATION
68 BLOWING ENGINE HOUSE #3
69 BOILER HOUSE #3
70 BLAST FURNACE "A"
71 BLAST FURNACE "B"
72 BLAST FURNACE "C"
73 BLAST FURNACE "F"
74 BLAST FURNACE "G"
75 BLAST FURNACE "H"
76 ORE LIMESTONE BINS
77 ORE BINS
78 COKE LIMESTONE BINS
79 ORE BINS
80 TRAVELING BOAT UNLOADERS
81 BOOSTER HOUSE
82 BENZOL BUILDING.
83 STORAGE TANKS
84 RAIL UNLOADER
85 BOILER HOUSE
86 CONDENSING & AMONIA HOUSE
87 SETTLING BASIN
88 CIRCULATING PUMP HOUSES
89 TAR STORAGE
90 CAR DUMPER
91 KOPPERS OVENS 3 BATTERIES
92 COKE WHARF
93 QUENCHING STATION (COKE DEPT)
94 SCREENING STATION
DELINEATED BY: FRANCISCO HIDALGO-ROMERO: -1986
BETHLEHEM STEEL PLANT BUFFALO 1924
LACKAWANNA, NEW YORK
OFFICE OF ARCHEOLOGY AND HISTORIC PRESERVATION
HERITAGE CONSERVATION AND RECREATION SERVICE
UNITED STATES DEPARTMENT OF THE INTERIOR
SHEET 2 of 4
HISTORIC AMERICAN ENGINEERING RECORD

there were ever-increasing numbers of shareholders to divide up the corporate pie so that earnings per share necessarily were diluted. The highest earnings-per-share level occurred very early in Bethlehem's history, then declined over the years as the number of shares increased. For example, in 1941-1947 Bethlehem had 5 million shares, but by 1950 there were 20 million, following a stock split. By 1960, after a second split, there were 80 million shares. When a stock splits, the face value of the stock — its market price — declines as well as the earnings, and theoretically this should not dilute the earnings per share. Nevertheless, the sheer numbers of shares outstanding represented an increased obligation on company coffers.

As we can see, despite the enormous vitality of the company's middle period, despite nearly 5 decades of quite steady growth and increased production, the company was already experiencing stress as early as the immediate post-war period. Regardless of its strengths as one of the nation's paramount steel companies, problems already existed which would lead the corporation into serious difficulty in its last years.

5. THE PLANT

The 1920s

After Bethlehem Steel Corporation acquired the Lackawanna steel plant in 1922, it initiated a comprehensive modernization program. In the two decades that had passed since Lackawanna Steel uprooted its operations from Scranton, piecemeal modifications in iron and steel technology had rendered some of the original installations obsolete while accelerated wartime operations had also taken a toll. Though Bethlehem began upgrading its recently-acquired Cambria and Midvale plants in Pennsylvania at about the same time, the New York plant was assigned a generous transfusion of new capital expenditures, commensurate with its stature as the logical point for serving the growing Midwest under the corporation's revised marketing strategy. Bethlehem President Eugene G. Grace observed that water delivery of finished products from Lackawanna to Great Lakes ports was one of the plant's natural advantages which the previous ownership had not exploited to the fullest. Lackawanna's strategic geographic position for raw material assembly and product delivery by lake, canal, and rail remained as attractive to Bethlehem as it had been to Lackawanna Steel's investors. Principal improvements included new coke ovens, larger gas blowing engines, revamped open hearth furnaces, and major additions to the plant's capacity for rolling structural steel.

Coke Ovens

Lackawanna's original by-product coke ovens fell under the axe of modernization. The 5 Otto-Hoffman batteries vanished into the dustbin of history by 1925, and the 5 Rothberg batteries followed by 1930. In their place Bethlehem constructed Koppers-Becker ovens with vertical gun-flue heating arrangements. Gas was burned on an entire wall simultaneously rather than alternately heating half of each wall as in earlier Koppers designs. Combustion products passed over the top of the coking chambers and then down to the appropriate regenerator. Ovens of this type minimized harmful pressure differences and cross-flow of gases between the flues and ovens and between adjacent regenerator walls.

Two new batteries, each with 57 Koppers-Becker ovens, began operating on 3 September 1924, situated just south of the 1920 Semet-Solvay battery. The ovens operated on a cycle of 11½ hours between charges, each oven taking a coal charge of 12 tons and producing 2100 lbs./hr. By comparison, the Semet-Solvay battery was running on a schedule of 16½ hours, with a 15-ton charge and an output per oven of 1800 lbs./hr. With the addition of 57 more ovens at a third Koppers-Becker

Plan of the Lackawanna plant, Bethlehem Steel Corp., as of 1924.

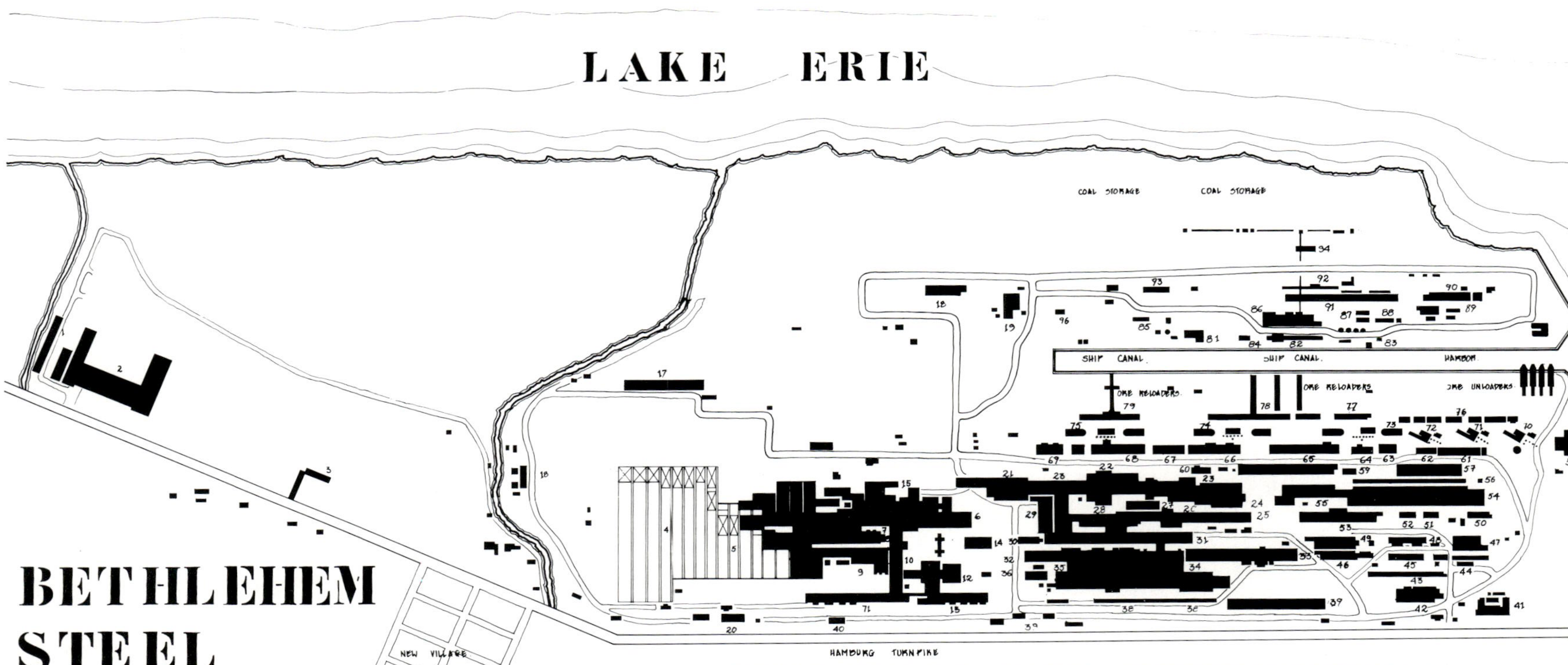

BETHLEHEM STEEL PLANT BUFFALO 1930

SCALE

1 BRICK SHED #1-2	28 BILLET MILL #8	55 BESSEMER DEPARTMENT	82 BENZOL BUILDING
2 NARROW GAUGE LOC. HOUSE	29 TWO CRANE RUNWAYS.	56 GAS PRODUCER DEPART #2	83 STORAGE TANKS
3 PATTERN STORAGE	30 PHYSICAL LABORATORY SHIPPING	57 STOCK YARD OH. DEPT #2	84 RAIL UNLOADER
4 STRUCTURAL SHIPPING YARD	31 BAR MILL #8	58 POWER STATION #2	85 BOILER HOUSE
5 RUNWAYS 1-10. RUNWAYS 10-15	32 BAR MILL #6	59 POWER HOUSE #1	86 CONDENSING & AMMONIA HOUSE
6 MILL BUILDING #15	33 BAR MILL #10	60 N.G. LOCOMOTIVE HOUSE	87 SETTLING BASIN
7 MOTOR ROOM #12 & #15 MILL	34 OPEN HEARTH #1	61 BLOWING ENGINE HOUSE #4	88 CIRCULATING PUMP HOUSES
8 MILL BUILDING #12	35 COVERED MOULD YARD O.H.	62 LADLE HOUSE	89 TAR STORAGE
9 MILL BUILDING #14	36 STRIPPER BUILDING	63 BOILER HOUSE #1	90 CAR DUMPER
10 BILLET YARD #12	37 STOCK YARD DEPT. #1	64 BRICK STORAGE	91 KOPPERS OVENS 3 BATTERIES
11 SHEAR BUILDING	38 GAS PRODUCERS TAR STORAGE	65 BLOWING ENGINE HOUSE #2	92 COKE WHARF
12 PLATE MILL #5	39 OFFICES YARD DEPT #1	66 BOILER HOUSE #2	93 QUENCHING STATION (COKE DEPT)
13 SLABBING MILL #4	40 STOREHOUSE	67 PUMPING STATION	94 SCREENING STATION
14 BILLET YARD	41 NORTH OFFICE PLANT	68 BLOWING ENGINE HOUSE #3	
15 STRUCTURAL MILLS #2-#3	42 STRUCTURAL REPAIR SHOP	69 BOILER HOUSE #3	
16 PIG CASTING MACHINES #4-#5	43 ROLL SHOP	70 BLAST FURNACE "A"	
17 SCRAP RECLAIMING PLANT	44 STORE HOUSE	71 BLAST FURNACE "B"	
18 THAW HOUSE	45 PIPE TOOL SHOP	72 BLAST FURNACE "C"	
19 OFFICE: CALCINING PLANT	46 BLACKSMITH'S / FORGE SHOP	73 BLAST FURNACE "F"	
20 CAR REPAIR SHOPS	47 ELECTRIC REPAIR	74 BLAST FURNACE "G"	
21 COLD SAW HOUSE	48 BOILER SHOP / PLATE SHOP	75 BLAST FURNACE "H"	
22 RAIL MILL #1	49 MECHANICAL REPAIR SHOP	76 ORE LIMESTONE BINS	
23 DROP TEST MILL #1	50 CARPENTER SHOP	77 ORE BINS	
24 HOT SCALE ON NG.	51 PATTERN SHOP	78 COKE LIMESTONE BINS	
25 44" BLOOMING/ RAIL MILL	52 PATTERN STORAGE	79 ORE BINS	
26 BLOOMING MILL #7	53 FOUNDRY	80 TRAVELING BOAT UNLOADERS	
27 BILLET YARD	54 OPEN HEARTH DEPART #2	81 BOOSTER HOUSE	

DELINEATED BY FRANCISCO HIDALGO-ROMERO: 1986

BETHLEHEM · STEEL · PLANT · BUFFALO · 1930
LACKAWANNA, NEW YORK.

HISTORIC AMERICAN ENGINEERING RECORD
NATIONAL PARK SERVICE
UNITED STATES DEPARTMENT OF THE INTERIOR

SHEET 3 of 4

HISTORIC AMERICAN ENGINEERING RECORD

battery, annual coke capacity reached 1,344,000 net tons by 1930, nearly double the 1920 capacity and equal to the maximum that the plant's blast furnaces would require if operating at full output. The Semet-Solvay ovens eventually became known as #1 battery, the Koppers-Becker batteries as #2-#4.

Other facilities in the coke oven department were also rebuilt or renovated. New coal handling equipment included a barrel-type car dumper, capable of emptying a 120-ton railroad car; a new coal handling and storage facility appeared between the ovens and the lake. Coke screening facilities were improved as were considerable portions of the system that reclaimed by-products for such coal-based chemicals as fertilizer, motor fuels, and synthetic dyes. Bethlehem engineers also increased the application of coke oven gas as fuel for the open hearths and rolling mill soaking pits or heating furnaces, replacing many of the coal-burning gas producers and adding another strand of plant-wide energy transfer that integrated operating departments even more tightly.

Blast Furnaces

Lackawanna Steel's Koerting gas blowing engines met the same fate as the original coke ovens. As chunks of the historic Koertings headed for the scrapyard, new gas blowing engines designed by Bethlehem took their places. Longer than the original models, the horizontal Bethlehem engines were squeezed at angles into the original blowing engine houses. The new units continued to recycle waste blast furnace gas as fuel. A sparkplug ignited a gas-air mixture inside a cylinder (much like an automobile); the resulting explosion moved a piston that forced air from adjacent cylinders into the cold blast mains and toward the furnaces. The Bethlehem engines provided more blowing power than the Koertings while consuming an equivalent amount of gas. It had taken three of the old engines to furnish the blast for a single furnace; when the first five Bethlehem engines went to work in the southern portion of Blowing Engine House #3, they moved enough air for two furnaces and half that required for a third stack. A secondary washing system removed residual dust and other sediment from the gas to prevent fouling of the engine parts. By 1930 all the Koerting engines had been retired, and the plant's complement of blowing engines consisted of 15 horizontal gas engines in use with five more under construction; six steam reciprocating double vertical engines also remained in place. The entire set of blowing engines could deliver 529,000 cubic feet of air per minute at 15-25 psi.

Beginning in the 1920s Bethlehem Steel also carried out a demolition and reconstruction program on the blast furnaces themselves. In the process Lackawanna Steel's original practice of numbering the furnaces was changed to a letter identification system, starting with A Furnace, the northernmost, and proceeding south. Bethlehem inherited a total of 9 blast furnaces from Lackawanna Steel. The renovation program progressed as follows:

Bethlehem Furnace	Lackawanna Number	First Operated	1920s Status
A	9	1918	Relined 1925
B	8	1918	Relined 1927
C	7	1907	Relined 1926
D	1	1903	Abandoned 1928
E	2	1903	Abandoned 1928
F	3	1904	Rebuilt 1929
G	4	1905	Relined 1925
H	5	1905	Rebuilt 1927
J	6	1906	Rebuilt 1928

The principal casualties of initial modernization were the plant's two smallest and oldest furnaces, the antiques from Scranton. On the eve of the Great Depression total pig iron capacity of the seven remaining furnaces amounted to 1,620,000 tons.

Bethlehem also initiated improvements in ore handling equipment, ordering a new Hullett-type unloader in 1928 as a first step toward phasing out the original machines. Specifications for the new model included 17-ton bucket capacity and

Plan of the Lackawanna plant, Bethelehem Steel Corp., as of 1930.

15
14

electrical operation throughout, replacing the hybrid electrical and hydraulic systems of the earlier models.

Steelmaking

Investments in the steel departments centered on #1 open hearth shop. The six oldest furnaces, 60-ton stationary models built during 1903-1904, were replaced by six 100-ton furnaces designed to operate on coke oven gas or tar instead of producer gas. The remaining eight furnaces in #1 shop continued to use gas from the nearby producer house. In conjunction with the new furnaces the overall structure of the building was strengthened; raw material and product handling capabilities were upgraded. A new stockyard with four 10-ton cranes replaced the existing cramped quarters where scrap and other solid materials were loaded into charging boxes. A banked trestle was built to connect the stockyard with the stock platform so that a train of scrap buggies could be run up directly to the level of the charging floor; previously, some charging boxes were hoisted up one-by-one to the charging floor level.

For handling heavier hot metal charges a 75-ton crane was erected. A 165-ton ladle crane was installed on the opposite side of the furnaces where molten steel was tapped. Also added were 3 new pouring platforms where steel was teemed from the ladle into ingot molds. Safety features taken into consideration during the rebuilding program included moving the operators' cages on the ladle cranes from the side nearest the furnace to the pouring side as well as providing an escape platform for the operators in case of a spill. Fine-tuning improvements on all the stationary furnaces in #1 and #2 shops included installing electric motors to replace hydraulic door hoists that were difficult to operate in cold weather. As of 1930 Lackawanna's ingot capacity was pegged at 1,980,000 tons, including 270,000 from the duplex process.

BLAST FURNACES
Ore docks. Hullett unloader with operator in grab bucket leg lifts iron ore out of lake vessel (1940).

Rolling Mills

Modernization in the rolling mills proceeded in two directions. First, several new structural mills were constructed as Lackawanna became Bethlehem's western base for penetrating markets around the lakes. Second, a number of existing mills were extensively revamped with a heavy emphasis on conversion from steam engine to electric motor drive.

Three new structural mills were laid down at Lackawanna during the 1920s. First to be completed at the end of 1925 was the 28" mill, known as #12. It superseded Lackawanna Steel's 1904 #3 mill and physically replaced the #11 16" bar mill. The 28" was designed to produce I-beams up to 24", channels up to 15", and angles up to 8" x 8" as well as sheet piling which Lackawanna Steel had introduced in 1908. Equipment in the new mill consisted of four sand-bottom regenerative batch heating furnaces fueled by coke oven gas, one 2-high stand of 35" reversing rolls with electric manipulators, and three 28" stands in line. The first two 28" stands featured 3-high rolls with traveling tilting tables and roller trains that moved the steel up or down as well as back and forth, since rolling occurred on two horizontal planes in 3-high mills. Electrical equipment in the mill included a 3500 hp motor driving the 35" rolls and a 5000 hp motor for the 28" stands. Original finishing equipment included a roller straightener, two gag straightening presses for I-beams, a punch for sheet piling, two cold saws, and one angle shear. Products rolled in the 28" were delivered to the structural mill group's new shipping yard which originally contained six bays.

BLAST FURNACES
Bethlehem horizontal gas blowing engines. Installed in the 1920s, these engines were superseded by turbo-blowers. Some veteran workers said that blast furnaces made better iron with the pulsating air blast from the reciprocating engines than with the constant blow supplied by turbos.

The 14" mill, completed in 1926, made light beams, channels, and angles as well as flats and rounds. Mill equipment included two continuous heating furnaces fired by coke oven gas and ten 2-high roll stands. Five stands of 18" rolls operated on the continuous principle. The 14" stands featured Skew-Y tables for handling long bars between stands 7, 8, and 9. Transfer tables changed the direction of the steel after passes in stands 6 and 9. As in the 28" mill, electric motors furnished

the drive for the roll stands. The 14" also included double hot beds, roller straighteners, and runout tables to the shears.

In 1927 Bethlehem rounded out the structural line at Lackawanna with mills for rolling its trademark Grey beams. As of 1897 Henry Grey had devised a means of rolling wide-flange beam sections on universal mills with narrow, ungrooved horizontal and vertical rolls. The Grey beams were equal in strength to existing structurals but lighter in section; fabrication was simplified since rolling the shape as a single section eliminated the need for building up wide-flange beams by riveting smaller structural elements together. Grey's original mills in Luxembourg and Duluth were not crowned with overwhelming success, but his design became the foundation of Bethlehem Steel's Saucon Division in Pennsylvania and eventually served as the corporate logo. Bethlehem had the largest structural capacity in the U.S. by 1920. It dominated the market for structurals in the east and planned the extensions at Lackawanna with a view toward similar success in the Midwest. Lackawanna's Grey mills consisted of a 54" 2-high 1-stand reversing blooming mill operating in conjunction with four 48" 2-high stands. This 54"-48" mill became known as #15.

These three new mills raised Lackawanna's structural steel capacity from 150,000 tons as of 1920 to over 950,000 tons. As a percentage of the plant's finished hot rolled products, structurals increased from 16.8% in 1920 to 39.3% in 1930. Higher production went hand-in-hand with natural marketing advantages. Grace's vision of products delivered by water from Lackawanna was fulfilled in 1928 when weekly boat shipments to Chicago and Milwaukee commenced. In that same year Lackawanna shipped all the structural steel required for Chicago's new Merchandise Mart direct to the riverside construction site.

BLAST FURNACES
Detail of gas blowing engine #7 in #2 Blowing Engine House. (L) starter jack. (C) flywheel. The jack was used to move the flywheel into starting position (1985).

In addition to building new structural capacity, Bethlehem invested substantially in production equipment for several existing mills. The #7 40" blooming mill was rebuilt with a completely new mill train. The soaking pits were rebuilt to use coke oven gas, and two new 15-ton stiff-leg pit cranes were installed to handle larger ingots. New electric motors drove the rollers on new approach and delivery tables that carried the steel into and out of the mill stand. The 2-high stand of reversing rolls that reduced the ingot into a bloom was now driven by a 7000 hp motor. A new 800-ton hydraulic shear cropped the ends off the blooms. The hydraulic manipulator that turned the bloom between passes was electrified. A mechanical pusher, reciprocating transfer, and piler handled the blooms for delivery to the billet mill. At the #5 48" universal plate mill, electric drive in the form of a 5000 hp motor also replaced steam power, and electrification likewise superseded pneumatic door hoists on the heating furnaces and also on the hydraulic slab pusher.

The production technology of the original Lackawanna plant had represented the crystallization of major changes in iron and steel engineering around the turn of the century. The extent to which subsequent incremental changes modified technology and work environments may be followed most closely in the 32" rail mill, the jewel of the Lackawanna complex when it was first laid out.

Lackawanna Steel had introduced two major modifications in its process of rolling rails. Around 1914 it added a large machine with milling cutters which removed partially decarburized steel from areas of the rail head and base where seams had led to eventual breakage. The deseaming operation took place between the existing roughing and finishing stands. In 1921, its last year of independent operation, Lackawanna Steel tore out the steam engines for the finishing stands and replaced them with a 7000 hp electric motor. The new equipment represented an early example of using a large reversing DC motor for rolling mill drive, following the precedent established at the 36" universal plate mill of U.S. Steel's South Works in Chicago. However, in Lackawanna's rail mill, the reversing capability could not be utilized to the fullest extent until the remaining stands were converted to similar motors.

These developments and several others were implemented when Bethlehem Steel reconfigured the 32" mill during 1928-1929. The rail mill department was not the primary focus of Bethlehem's original plans since other areas were determined to be in need of more immediate attention. However, continuing changes in customer

BLAST FURNACES
Skip car on G Furnace for charging raw materials (1985).

5-TON.
11738
10-15-31

rail specifications and operating practices were pressing up against the limits of the original design and capacity.

Bethlehem's modernization program was concentrated in the rail mill itself, the finishing areas being left basically unchanged. The first three stands of the 1903 mill, which handled the blooming passes, were eliminated along with the original soaking pits. The rail mill subsequently received blooms from a new 44" mill, built to handle the larger ingots from the open hearths that had supplanted the original Bessemer converters. Rather than rolling the rail directly from the ingot, the sheared blooms were now reheated in two flat-bottom hearth-type regenerative batch heating furnaces. Four-ton charging and drawing machines moved the blooms in and out of the 16 doors on the reheating furnaces. The rump of the rail mill now consisted of a single 36" 2-high roughing stand and the two original 32" finishing stands. Bethlehem also rounded out the conversion from steam to electric drive by installing a 5000 hp reversing motor on the roughing stand.

The modernization program of the 1920s thus laid down the basic equipment that the 32" mill crews used for rolling rails, billets, and other sections over the next half century. Many of the refinements that Bethlehem introduced during its extensive upgrading had similar consequences for the relationship between technology and work. Substituting one form of power or mechanism for an earlier version lowered operating costs without necessarily displacing workers or transforming work processes to the same extent as the engineering innovations originally installed by Lackawanna Steel. As the relative pace of technological change slowed and the job structure stabilized, one of the preconditions for industrial unionism as a new stage in labor relations began to emerge.

Depression and War, 1930-1945

BLAST FURNACES
Cast house with runner for molten iron in floor (1956).

The successive collapses of the international economy and international peace were felt inside the Lackawanna steel plant as well as around the world. The onset of the Depression strangled Bethlehem's modernization program, halting construction of new open hearth furnaces which had gotten underway in 1929. Reduced demand for steel subsequently retarded the overall pace of plant expansion. Nevertheless, during the mid-1930s Bethlehem did create a facility of enormous significance for the future of the Lackawanna plant: the hot and cold strip mills. This complex marked the first major departure from Lackawanna Steel's original plant layout, and its construction during the Depression indicated the growing importance of flat-rolled products for consumer durable goods in the nation's steel industry. As war again engulfed Europe for the second time in a generation, steel output rose, and the federal government began financing additions to plant capacity.

Coke Ovens

Coke oven and by-product recovery capacity remained static throughout the 1930s. At the end of 1940 Bethlehem announced a program of expansion for war production that included a new battery of ovens and additional toluol by-product capability at Lackawanna. The new #5 battery began operating in August, 1941. Construction on another new battery, #6, was underway by April, 1942. Both batteries contained 76 Wilputte ovens with gun-flue heating systems. Like the existing Koppers-Becker ovens, the Wilputte design featured vertical heating flues but with a different arrangement for obtaining even heating of the coke from end to end in the retort oven. The Koppers-Becker ovens had burned gas mixed with preheated air the entire length of alternate walls. The walls of the Wilputte oven were divided into zones. Combustion and the passage of hot waste gases to the regenerators alternated between two outer zones and a double inner zone rather than from wall to wall. The three extant Koppers-Becker batteries were also rebuilt during the war years. By 1945 the annual capacity of Lackawanna's six batteries totaled approximately 2,000,000 tons of coke, about a third greater than in 1930.

STEELMAKING
Open hearth charging floor with 5-ton charging machine and 75-ton overhead charging crane (1931).

11735
10-6-31

Blast Furnaces

At Lackawanna the Depression years were basically a holding period in the blast furnaces as well. G Furnace was rebuilt in 1931 and again in 1938; the old C and H furnaces were dismantled. However, the prospect of war provided a stimulus for expansion in this department also. The steel industry and the government both sought to augment existing pig iron capacity. They anticipated that scrap steel would be in short supply and that open hearths would therefore have to use higher molten charges. Thus, more blast furnace capacity would be required along with greater coke capacity to fuel the new stacks.

As a result of these projections, two new blast furnaces were built at Lackawanna: H furnace in 1941 and C furnace in 1943. At 103' 8" these were the tallest stacks at the plant to date and were designed to produce 1200 tons of pig iron per day, four times the capacity of Lackawanna Steel's first furnaces. This scale of daily output required charging 2160 tons of ore, 996 tons of coke, and 384 tons of limestone flux into H furnace. Skip cars with a capacity of 220 cubic feet handled the charging of this burden; new control equipment was designed to insure that the raw materials were charged in proper sequence and that variations in the height of the stock line were automatically recorded. When it first began operating, the new H furnace used the stoves from Lackawanna Steel's former #5 stack. H furnace was also the first at Lackawanna where the blast was supplied by a turbo-blower rather than reciprocating steam or gas engines. During August, 1942 H furnace broke production records by averaging 1440 tons per day. By the end of the Second World War the aggregated annual capacity of the two new and five older furnaces was 2,550,000 tons of pig iron, 36.5% greater than 1930 and 43.6% higher than 1938.

STEELMAKING
Charging machine operator empties boxes of materials (R) into open hearth furnace (1931).

Steelmaking

In September of 1929 Bethlehem Steel announced plans to construct a third open hearth shop with eight furnaces. Initiated at an unpropitious moment with the stock market crash just around the corner, the project remained unfinished for several years. However, work was resumed during August, 1936, and six 150-ton furnaces went into operation during 1937. The additional open hearths were needed to supply the new strip mills.

During the same period in the mid-1930s Lackawanna's Bessemer converters and duplexing capacity were finally phased out. With the passing of the Bessemers the flames that had been heralded as the dawn of a new industrial age in Buffalo dwindled away. With the completion of #3 open hearth, Lackawanna's basic open hearth steelmaking capacity in 1938 stood at 2,592,000 tons from 30 furnaces fired by a combination of coke oven gas, producer gas, tar, and oil. Toward the end of the war the capacity of the same number of furnaces was reported as 3,120,000 tons.

Rolling Mills

During the period between the two world wars, one of the salient trends in the steel industry was the growth in demand for flat-rolled products, sheets and strips whose width was much greater than their thickness. Manufacturers of automobiles, appliances, office furniture, and containers consumed 1,358,000 tons of flat-rolled steel in 1911 and 7,823,000 tons by 1935. By contrast, sales in such staple lines as rails and heavy structurals plummeted along with nationwide capital expenditures during the early years of the Depression. Large integrated plants such as Lackawanna therefore faced the task of rounding out their product lines with flat-rolled capacity and incorporating the new mill technology into existing layouts.

Until the mid-1920s steel sheets were rolled primarily on single-stand mills. Some of those mills had been mechanized in the early 20th century with tilting tables replacing men and their tongs in handling hot steel. Other mills continued to be

STEELMAKING
Checking charging boxes in open hearth. (L) iron ore. (R) limestone. Charging machine in background (1931).

hand operated by teams of skilled workers. Where sheet rolling had survived as a manual operation, the production process remained comparable in many ways to the 19th-century iron and steel industry prior to consolidation and mechanization. The stamina and knowledge of workers continued to be important factors in daily production, and occupational ecology at the smaller sheet mills nourished some of the last craft strongholds of the Amalgamated Association of Iron, Steel and Tin Workers. As demands for flat-rolled steel began accumulating, the productivity of a process that could be dominated by labor came into question. However, variations in size, texture, and composition of hand-rolled sheets posed even more serious issues. Sheet supplies with uniform characteristics became critical elements in mass-production stamping technology, particularly when automobile manufacturers began using more steel in car bodies.

The steel industry's eventual solution to flat-rolled output and quality concerns was a rolling mill that produced longer, wider strips by passing the steel continuously through separate stands of ungrooved rolls lined up in tandem, rather than rolling shorter sheets on individual stands. Several related engineering developments made the continuous hot strip mill a technical reality. Four-high mill stands where two work rolls of smaller diameter were supported above and below by two larger backup rolls provided a means of exerting the greater pressures required in reducing strip steel to desired gauge. Convex work roll surfaces prevented the steel from slipping from side to side as it passed through the finishing stands at progressively greater speed. Hot coilers wound the mill's product up in a convenient package for further processing or shipping. A number of these features were devised in mills at Ashland, Kentucky which rolled sheets in multiple lengths as of 1924, and Butler, Pennsylvania where strips were rolled in coils as of 1926; the American Rolling Mill Company eventually consolidated control over the patented elements of hot strip mill design.

STEELMAKING
Charging hot metal into open hearth with overhead crane. The crane operator tilts the ladle by raising one side with an auxiliary hoist.

Large continuous hot strip mills provided manufacturers with long, wide, uniform sheets and strips. A complementary cold-rolling process permitted even greater reductions in thickness while simultaneously improving the surface and mechanical properties of the steel used for auto body stampings or other mass-production industries. Combined hot strip and cold reduction mills first began operating at Inland Steel's Indiana Harbor works in 1932. The new strip mills required massive capital investments, and their introduction altered the geography of sheet production along with its work process. Smaller hand mills scattered throughout the Ohio Valley began withering away, and production clustered instead around the integrated plants of the larger companies that could afford to install the strip mills.

At Lackawanna land was still available for expansion; as a result, in 1935 Bethlehem designated Lackawanna as the site of the corporation's first strip mill. Prior to the 1930s, there had been no sheet mills at all within the plant, although the old #8 18" mill had supplied semi-finished sheet bars to the Seneca Iron and Steel Company of Blasdell since 1907. In 1932 Bethlehem acquired the Seneca plant and its 150,000-ton capacity for hot-rolled and cold-finished automotive and furniture sheets; the corporation's only other sheet mills were then located at Sparrows Point, Maryland. Absorbing the Blasdell operation gave Bethlehem quick entry into the sheet market at a time when sales in heavier product lines were particularly slow. By 1935 125,000 tons of Seneca sheets made up 7.6% of Lackawanna's finished hot-rolled products. However, sheet production on this limited scale represented only a temporary expedient. It would take the strip mills to make a major impact on both the market and plant production.

Ground was broken for the construction of the strip mill complex in March, 1935 on a 67-acre tract formerly used for gardens by plant employees. The site chosen lay across the Hamburg Turnpike southeast of the area originally occupied by Lackawanna Steel. The hot strip mill was ready for operation as of 2 January 1936. The annual capacity of the hot strip and cold reduction mills was pegged initially at 600,000 and 250,000 tons respectively.

The basic components of the 79" hot strip mill when it first opened included slab reheating furnaces, a continuous roughing train of four stands, and finishing train of six stands, plus equipment for disposing of coils, sheets, and light plates.

STEELMAKING
Teeming ingots, #1 open hearth (1951).

12

Slabs up to 5" thick, 50" wide, and 15' long were delivered to a storage area. Three recuperative reheating furnaces, each rated at 55 tons per hour, delivered slabs onto the roller line at 2250°. The slab first passed through a 2-high scale breaker to remove its accumulated coating of iron oxide acquired in the furnace, before arriving at the first roughing stand. Number 1 rougher was a 4-high stand with 36" diameter work rolls and 49" diameter backup rolls, all 96" long. Slabs for sheets or strip less than 48" wide passed straight through; however, greater widths required cross-rolling where a turntable manipulated the slab so that it could be pushed through the rolls broadside. Cross-rolling thus allowed mills to produce strip of considerably greater width than the widest slab available. After the broadside pass, a second turntable returned the slab to its original direction, and a slab squeezer squared up its edges within the required tolerances for width while evening its thickness. The roughing passes were completed in #2-#4 stands, each having 4-high rolls of 2½" and 49" diameters x 79" long as well as vertical edging rolls. After the roughing passes, a 4½"-5" thick slab was reduced to ⅝"-1¼". Since rolling temperature influenced the grain structure and quality of the final product, an intermediate roller table between the roughing and finishing trains allowed the heat of the steel to be controlled by water and air-cooling devices. At 1900-1950° the slab then entered a second scale-breaker, had its ends cropped in a flying shear, and then passed through six 4-high finishing stands closely grouped in tandem. Roll dimensions on the finishing stands were identical to #2-#4 roughers. Loopers regulated the tension in the strip during the finishing passes. The strip exited from the mill at speeds between 676 and 1352 feet per minute, being sprayed with water to quickly cool it below the required finishing temperature, thereby improving its metallurgical properties. High-pressure water sprays or steam removed scale from the strip at several points in the mill. Thus, in a consecutive sequence of passes the hot strip mill could transform a slab into a strip 600' long, up to 72" wide, and as thin as 0.005 inches. Some of the product was then wound up like a roll of paper towels by two hot coilers for delivery to storage or the cold strip mills. The remainder was sheared for finishing as sheet or plates. Operations in the adjacent cold-rolling department included pickling, cold reduction, heat treatment, and finishing.

The strip mill complex embodied the close coordination of mechanical and electrical engineering that marked continuous sheet mills as the American steel industry's most significant innovation during the first half of the 20th century. Power was required along with precision. In the hot mill the four roughing stands were driven by 3000 hp constant speed AC motors, while the finishing stands were driven by adjustable speed DC motors: 3500 hp on #5-#7, 4500 hp on #8-#9, and 2500 hp on #10. Adjustable voltage control on the finishing train drives facilitated proper correlation of speed and cross-sectional reduction in successive passes.

Detroit swallowed 75% of the new strip mills' output. Hot rolled steel went into automobile frames while cold-rolled output was transformed into automobile fenders or hoods. Strips from the hot mill also were shipped to Bethlehem's tin mills at Sparrows Point. By 1938 hot rolled strip had jumped to 36.8% of the plant's finished hot rolled capacity, and cold rolled sheet capacity was four times higher than in 1935. In 1939 the 79" hot mill set a company record for an 8-hour shift by turning out 1845 tons.

However, for a time, older production processes continued in the shadow of the strip mills. Bethlehem maintained the old Seneca plant as a standby facility and then converted it to a galvanizing operation. Some of the sheets to be galvanized were rolled on updated 2-high mills with mechanized feeder and catcher tables while other Seneca mills still operated by manual methods. Thus, for a time Lackawanna workers might have encountered 19th-century handsheet technology as well as one of the most modern strip mill installations in the industry.

Though the hot strip and cold reduction mills represented by far the most spectacular development of the Depression era at Lackawanna, there were changes in other mills as well. In 1930 the #8 18" continuous billet and sheet bar mill was replaced by a larger 30"-21" billet mill. The 30" mill consisted of 6 stands of 2-high rolls; originally there were 6 continuous stands of 2-high rolls in the 21" train. The new mill continued to work on blooms direct from the 40" mill. In 1940 a new 12"

STEELMAKING
Adding recarburizing materials to ladle to regulate carbon content of finished steel as heat is tapped from open hearth furnace (1956).

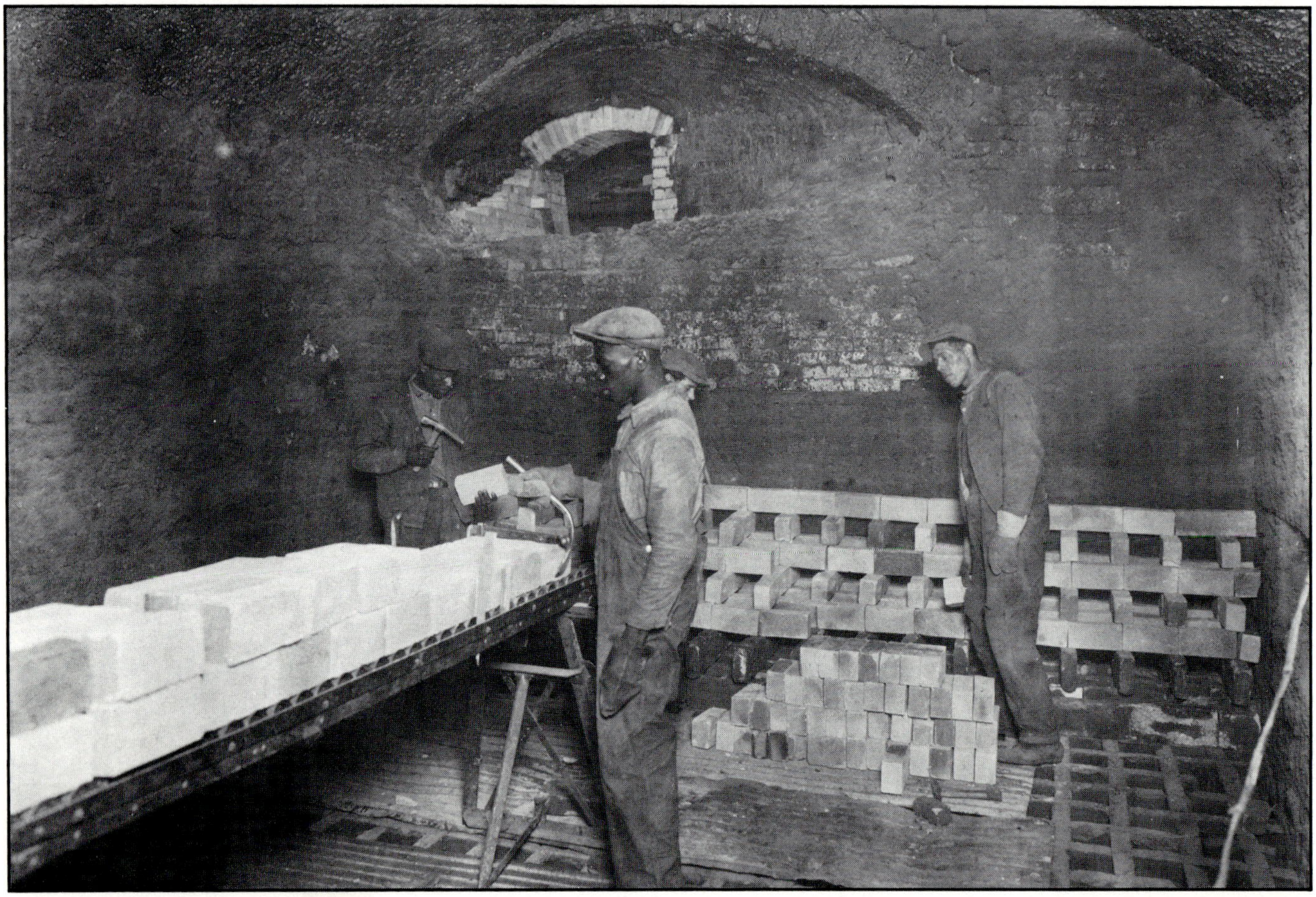

bar mill was laid out in the Town of Hamburg south of the strip mill complex. It replaced the original Lackawanna Steel #6 mill where limited capacity in the finishing department had become a bottleneck. Some of the roll stands from the 1905 mill were incorporated into the new facility, blending with two new continuous heating furnaces, a modern 275' hot bed, and equipment designed to control the quality of the steel throughout the process. The new 12" mill was designed to roll bars from 7/8" to 2½" diameter or squares up to 2". Initial orders included shell rounds and auto forgings stock. The new bar mill replaced an outmoded facility, but its location, combined with the strip mills, shifted the plant's center of gravity even further away from the strict rationality of Lackawanna Steel's original layout. Integrating the flow of materials, products, and energy through the plant thus became more difficult with the dispersal of the new mills to more distant areas.

The Long Boom, 1946-1962

Through the years of U.S. domestic prosperity and international economic dominance that followed the Second World War, the growth of the Lackawanna Steel plant was contingent on allocations from Bethlehem's capital expenditure budget. The increase in plant capacity over this period reflected the continuing significance of the factors that first had brought steelmaking to the Buffalo area, including an advantageous location for assembling raw materials, making iron and steel at comparatively low cost, and distributing products to growing markets. Lackawanna's value in Bethlehem's corporate program was particularly apparent during Edmund F. Martin's tenure as the plant's general manager (1950-1958).

STEELMAKING
Dolomite-throwing machines began replacing manual labor during the 1920s to repair open hearth furnace linings.

War production had depleted the nation's reserves of raw materials, particularly the iron ores of the Mesabi Range. However, the steel industry's experiments with beneficiation of lower-grade ores such as taconite achieved commercial viability by the mid-1950s. The flow of raw material carried on the Great Lakes thus remained relatively intact, though organization of similar extraction and transportation systems on a global scale would eventually aid steelmaking sites at coastal locations as well. Opening of the St. Lawrence Seaway in 1959 was expected to improve Lackawanna's access to the recently-discovered ores of Labrador.

Investments in new equipment at Lackawanna during this period represented modifications and refinements of existing technology rather than dramatic innovations on the scale of the hot strip mill. Quality control assumed increasing importance, and changes at the various stages of production reflected pressure from customer specifications. Flat-rolled steel for consumer products became the keystone of plant operations, reversing the original mixture of light and heavy products at the turn of the century. Midwestern markets were served increasingly by truck as well as by water and rail shipments.

However, there were changes on the horizon that would affect the economics and technology of the steel industry in general and the position of the Lackawanna plant in particular. The shock of those changes would be magnified even more because they followed so closely on the prosperity and growth of the post-war era.

Coke Ovens

Additions to the coke plant included #7 battery in 1952 and #8 battery in 1962. Both batteries consisted of 76 Wilputte ovens, the first at the plant to have underjet heating systems as distinguished from the older horizontal-flue or gun-flue arrangements. The underjet heating system introduced fuel gas into each vertical flue through separate burner pipes equipped with an orifice and metering pin that permitted more precise control of gas flow and easier maintenance. Air for combustion was drawn from an enclosed basement and was therefore independent of the variations in wind velocity, direction, and temperature that could affect gun-flue heating systems. The new Wilputte ovens featured high-low burner construction, designed for better vertical distribution of heat; in this arrangement the height of

STEELMAKING
Bricklayers rebuilding checkerwork in open hearth regenerative chambers (1929).

the burners in adjacent flues alternated to prevent the overheating due to concentrating all the burners at the bottoms of the flues.

As of 1953 the plant received 3,750,000 tons of coal in railroad cars for use in coking operations. To cope with this level of shipments, new coal handling and storage facilities were constructed in the mid-1950s, including a mobile system for transporting coal to and from the stockpiles. Coke capacity increased from 2,000,000 tons with six batteries as of 1945 to 2,800,000 tons with eight batteries.

Blast Furnaces

Projected increases in Lackawanna's steel capacity during the early 1950s required beefing up the plant's hot metal output to supply more open hearths. An additional blast furnace was authorized by Bethlehem in 1951 and completed in 1952 (along with the new coke oven battery that furnished its fuel). At 105' 3" high with a 29' diameter hearth, J furnace was the plant's tallest and largest. It replaced a stack originally built in 1906 and rebuilt in 1928. Centrifugal turbo-blowers supplied an ever-increasing proportion of the air required by the blast furnaces. Two additional units were installed during the 1950s while five of the gas engines were taken out of services. As of 1960 the five turbo-blower units in service had an average capacity of 100,000 cubic feet per minute while the 13 active gas engines averaged 18,000 cfm apiece. Annual pig iron capacity increased from 2,550,000 tons in 1945 at the seven existing furnaces to 3,798,000 tons as of 1960 from seven stacks including the new J furnace.

By 1953 lake carriers were delivering 5,500,000 tons of ore and 1,700,000 tons of limestone to the plant annually. To handle the waterborne traffic more effectively, Bethlehem modernized the ship canal originally excavated for Lackawanna Steel, deepening the channel to 24-27' to accommodate the largest ore ships. A new belt conveyor system augmented the older Hullett unloaders in delivering raw materials to the blast furnaces.

A Dwight-Lloyd sintering plant went into operation in 1950. Sintering involved the agglomeration of particles that were previously too small or otherwise unsuited for charging into the blast furnace: fine lake ores, flue dust, and coke breeze. Sintering machines consisted of a continuous series of pallets upon which the mixed materials were ignited and burned, producing a strand of granular material that could be used in the furnaces along with coarse and medium ore. The 1950 Lackawanna sintering facility featured two 6'-wide strands and a grate area of 612 square feet per strand; gas fuel was used for ignition; the plant's annual capacity totalled 1,080,000 tons. In addition to improving the physical structure of the material, the agglomeration process also enhanced the iron content of the fine lake ores. Sintering was thus one aspect of the broad trend toward beneficiation of lower-grade ores following the wartime depletion of higher-grade reserves.

Steelmaking

Expansion of Lackawanna's steelmaking capacity at first proceeded along conventional lines. Four open-hearth furnaces with an average capacity of 250 tons were added to #3 shop during 1952-1953; a fifth new open-hearth was completed in 1957, raising the plant's total to 35. New construction and incremental modifications combined to increase annual capacity from 3,900,000 tons as of 1951 to 6,000,000 tons as of 1960. During this period Lackawanna ranked as the third largest steel plant in the U.S., behind Sparrows Point and U.S. Steel's Gary works.

Within the steel industry, however, novel open-hearth practices that would eventually be implemented at Lackawanna began appearing during the post-war era. The increasing availability of commercially-manufactured oxygen made its use in steelmaking more economical. Experiments during the late 1940s demonstrated that adding oxygen to the air normally used for combustion increased the output from existing open-hearths while saving fuel as well.

Oxygen as a combustion aid proved particularly effective in speeding up the melting of the solid materials charged into the furnace such as steel scrap or cold pig iron. However, reaping the full benefits of faster melting required proportional improvements in scrap-charging methods as the number of scrap boxes to be handled over a given period increased. Oxygen enrichment also disclosed a need for improved refractories; the process was hard on furnace linings and additional down time for repairs could cancel out gains due to faster melting. In addition to cutting the time required for initial scrap charging and melting, oxygen was also found to shorten the working period at the conclusion of a heat by accelerating the chemical reactions that controlled elimination of carbon and raised the temperature of the bath.

By the early 1960s refinements in materials handling, refractories, and methods of delivering oxygen had been applied to the open hearths at Lackawanna. Combined with improved charging equipment, oxygen applied through roof lances cut the time required for making steel by over an hour to a low of 7 hours 35 minutes from tap to tap. With improved operating times and increased scrap utilization capability, proponents of the open hearth felt that its continued viability was assured even with the advent of a more revolutionary steelmaking process, the basic oxygen furnace. Another new technology was also added to the open hearths as electrostatic precipitators were installed to control air pollution caused by dust and smoke particles.

Rolling Mills

Changes at the rolling mills during this period of general prosperity were unevenly distributed. New heating and rolling facilities coexisted with some technology that dated back to the 1920s or earlier. Aside from major hardware, an equally significant variety of mechanical procedures for conditioning semi-finished steel became key elements in maintaining product quality.

The intertwined concern for output and quality was evident in one of the first major post-war construction projects. Announced in 1945 and completed by 1947, the plan called for new facilities to replace the 1911 semi-continuous 10" bar mill with a continuous 10" mill. To centralize bar production more efficiently, the new mill was laid out adjacent to the 1940 12" mill with common storage, billet preparation, finishing, and shipping departments. Intended for supplying carbon and alloy steel bars to aircraft, automotive, and agricultural implement manufacturers, the new 10" mill was designed with alternate horizontal and vertical rolls to eliminate twisting of the bar between stands. When it began operating, the annual rolling capacity of the 10" continuous mill was three times greater than the semi-continuous mill which it replaced.

Improving the quality of bars also entailed new arrangements in the blooming and billet mills that rolled the semi-finished steel for the new mill. Bethlehem refurbished the 40" and 30"-21" mills with an eye toward concentating all billet production for the bar mills there, including the high-grade carbon and alloy billets presently rolled on the 44" and 32" mills. Three five-hole recuperative soaking pits, 20' x 7½', were added at the 40" mill, and a continuous hot scarfing machine was installed to remove surface defects from the bloom after rolling. On the 30" mill three of the six extant horizontal stands were replaced with vertical stands to eliminate twisting the billet between stands which could open seams and thereby affect finished quality. The speed of the screwdowns on the remaining horizontal stands was upgraded to insure proper control over the amount of reduction. A 60" hot saw was installed between the 30" and 21" trains to cut billets into suitable length, as was a new steam-operated flying shear after the final 21" stand.

During 1952-1956 upgrading at the primary 40", 44", and 54" blooming mills included construction of additional soaking pits that could accommodate larger ingots; the 1950s-era pits at the various blooming mills measured 25' x 8' while the dimensions of the pits built in the 1920s were 16' x 7'. The newer pits were fired continuously on the recuperative principle in the interest of heating the steel as uniformly as possible. The older regenerative pits tended to expose ingots to uneven

heating on opposite sides due to periodic reversal of the air and gas flows. Two additional stands were added to the 21" mill in 1954 along with a new flying shear.

Bethlehem showered new equipment on the strip mills almost continuously. During the late 1940s and 1950s two new slab reheating furnaces were placed in the 79" hot strip mill, and an eleventh stand with rolls 30" and 49" x 79" was added to the ten stands originally built in 1936.

At this time a fourth 4-high stand was added to each of the tandem cold reduction mills, the 54" in 1947 and the 75" in 1950. In 1949 work began on major enlargements of cold-rolled sheet and strip capacity to complement the new Ford stamping plant which opened in nearby Woodlawn in 1950. During the following decade additional investments were made in the pickling, annealing, cold finishing, and shearing areas. The 79" mill's annual capacity, which had been rated at 720,000 tons as of 1938, totalled 2,040,000 tons in 1951, 2,400,000 tons in 1954, and 2,700,000 as of 1957.

In 1961, as capacity and output peaked at the Lackawanna strip mill, Bethlehem completed a 45" x 90" universal slabbing mill with 27 recuperative soaking pits each 28" x 11¼". Besides upgrading slab rolling capacity, the new mill also represented an attempt to rebalance the internal flow of production within the plant. During the 1930s the strip mill complex had been sited a considerable distance from existing primary mills as well as being divided from the older portion of the plant by the Hamburg Turnpike. Such separations increased the costs associated with extensive intra-works transfer of materials and heat, though originally the initial profit gains associated with continuous sheet and strip rolling dwarfed the negative consequences of a more dispersed layout. However, the passage of time pointed up the advantage of closing some of the spatial gaps between the sites of steelmaking, slab rolling, and strip production. The 45" x 90" mill, located at the southern end of the portion of the plant lying west of the Hamburg Turnpike, was relatively close to #3 open hearth (as well as the future site of the BOF). In 1962 Bethlehem announced that the slabbing mill would be paired with a second-generation hot strip mill to be built on the opposite side of the Hamburg Turnpike from the 1936 facility; however, that project never got off the ground. The 45" x 90" mill subsequently was conscripted to roll slabs for Bethlehem's new hot strip mill at Burns Harbor, Indiana while strip production at Lackawanna remained in its original location.

Bethlehem did open what was then the largest hot-dip galvanizing plant in the U.S. at Lackawanna during 1962. It was capable of handling the heaviest and widest coils in the industry, 70,000 lbs. and 72" wide. Galvanizing involved coating steel with corrosion-resistant zinc, a process first undertaken locally at the old Seneca sheet plant. However, production in the 1962 facility was considerably more automated. The galvanizing plant provided another outlet for coils from the strip mills. Lackawanna's specialty products shop used galvanized steel extensively in the manufacture of roadside barriers.

Indian Summer, 1962-1970

Lackawanna's capacity and post-war production reached record levels in the 1960s. Bethehem's eventual introduction of the basic oxygen process in 1964 revolutionized steelmaking to the same extent as continuous strip mills had transformed rolling during the 1930s. Combined with residual open hearth operations, the new BOFs raised rated plant steelmaking capacity to an all-time high. Intensive utilization of installed capacity kept the plant humming at record levels. In 1966, 6,583,000 tons of steel were produced, with 6,319,000 tons in 1968, and 6,580,000 tons during 1969; coke oven and blast furnace operations strained to keep pace.

However, despite record production and employment, Lackawanna's future was not really secure. There was a new factor that affected how Bethlehem would weigh and balance future expansion. Lackawanna's Indian Summer was chilled by clouds from the west, from Bethlehem's newest plant at Burns Harbor, amid the sand dunes east of Chicago and Gary.

STEELMAKING
Installing BOF Vessel (1964).

Bethlehem Steel had wanted an outlet in the Chicago area since the 1920s; it sought access through two unsuccessful mergers in 1930-1931 and again in 1954-1958 with Youngstown Sheet & Tube which operated a plant at Indiana Harbor. Had either merger been consumated, it could have precipitated Lackawanna's downfall as Bethlehem's primary western plant at an earlier date. However, since Bethlehem was ultimately unable to acquire an existing plant near Chicago via merger, it eventually took an expensive alternative of building a completely new works on its own. This decision reflected the culmination of a century-long movement of steel production and consumption toward the Great Lakes and Midwest. Demand for steel in the western portion of the manufacturing belt between Detroit and Chicago was growing more rapidly than any other section in the U.S. Bethlehem's position in this important market had deteriorated steadily as other companies expanded their plants in the Chicago area while cutting back elsewhere. The days were long past when Lackawanna could ship steel structurals into Chicago for major construction projects.

Bethlehem began acquiring land on the southern shore of Lake Michigan in Porter County, Indiana as early as 1956; in December, 1962 the corporation announced its decision to build the Burns Harbor plant on 3300 scenic acres. The Burns Harbor site provided Bethlehem with the same opportunity as Lackawanna Steel had seized in 1899: a chance to install some of the latest technology from the ground up at a location favored by the changing economics of the steel industry. Unlike Lackawanna Steel, however, Bethlehem did not undertake construction of a fully-integrated plant at the outset of the project. Instead, its plan was to start with very modern finishing mills that could meet growing Midwestern demand for flat-rolled products. Mills for rolling plate, cold reduced sheet, and tinplate were operating by 1965, and an 80" hot strip mill opened in 1966. The steel for these mills was made in Lackawanna and rolled into slabs for shipment to Burns Harbor on 85-car freight trains.

Record operating levels at Lackawanna thus represented a transitional phase coinciding with temporary lack of full integration at Burns Harbor. This new facility took over much of the marketing area that had once been Lackawanna's territory. The Burns Harbor project also absorbed much of the capital that, under other circumstances, could have been invested in a thorough modernization of the older plant; even in this period capital costs per ton were much lower for extensions of existing works than for new construction on a greenfield site, though a new integrated plant could expect to achieve significant operating economies. As Bethlehem continued the investment pattern that eventually rounded out Burns Harbor's steelmaking capability and terminated Lackawanna's supporting position, the New York plant faced a new role in a changing overall corporate strategy that would determine its future.

Coke Ovens and Blast Furnaces

Production of coke and pig iron during this period reflected the pressures of running older equipment in conjunction with record operating levels and the introduction of BOF steelmaking technology. The BOFs, in conjunction with surviving open hearths, required larger quantities of hot metal for their molten charges, but due to the circumscribed nature of modernization at Lackawanna during Burns Harbor's adolescence, the pig and coke needed to meet these new demands had to be coaxed out of the installations already on the ground. A new battery of 76 Wilputte underjet ovens pushed its first coke only at the end of the decade. The ovens in this new #9 battery were 6 meters tall, higher than any others at the plant. The #9 battery replaced both the #1 Semet-Solvay battery with 60 ovens and the #2 Koppers-Becker battery of 57 ovens. However, some of the plant's other batteries were beginning to show their age and defects in original construction. During 1963-1964, H and J furnaces were modernized to meet anticipated BOF hot metal requirements. The work at both furnaces included new stove burners and fans to increase blast temperatures by 400° (20%) and thereby raise output.

Steelmaking

During the Indian Summer years the process of steelmaking at Lackawanna was radically transformed by the installation of basic oxygen furnaces. Part of an expansion program announced in mid-September, 1962, the new BOFs were the first to be built at a Bethlehem plant, though the technology had already been adopted at other U.S. plants as well as in Europe where the process had originated. The function of the basic oxygen process was the same as the earlier Bessemer and open-hearth innovations: oxidation of the carbon, manganese, phospherous, sulphur, and silicon originally present in pig iron so that undesirable impurities were carried off in the form of gases or slag. However, in the BOF the mechanism for initiating these chemical reactions represented a new point of departure from existing practice, and the speed at which the reactions proceeded was greatly accelerated, thus shortening the time required for producing a heat of steel.

The new process involved blowing oxygen onto the surface of molten iron bath via a vertical lance inserted through the top of a vessel lined with basic refractory bricks. The technique was introduced in the Austrian towns of Linz and Donawitz and became known as the L-D process. It was developed during 1949 under conditions where scrap was in short supply and the pig iron available for hot metal was unsuitable for existing pneumatic processes. McLouth Steel of Detroit installed the first L-D units in the U.S. during 1955.

The converters installed at Linz and Donawitz bore a superficial resemblance to Bessemer vessels; they looked like eggs that could be tilted frontwards or backwards on trunnions. However, the oxygen lance replaced the tuyeres for blowing the bath. The sequence of operations in the first BOFs involved charging scrap and hot metal, blowing oxygen into the vessel, and adding fluxing agents plus scrap to control slag formation and bath temperature; as in the Bessemer process, flames issuing from the mouth of the vessel marked the progress of the various chemical reactions and provided a means of visually controlling the length of the blow. However, the speed of the reactions in comparison with the open hearth eventually resulted in the introduction of computer controls to supplement or supplant human judgment.

Construction of Lackawanna's two-vessel BOF shop got underway on 5 March 1963, and the first heat was poured on 3 October 1964. Each vessel was 38' tall and could cook up a 250-ton heat in 40 minutes. Introduction of BOF steelmaking also required construction of three kilns for reducing limestone into burnt lime since the ordinary flux alone did not act quickly enough to form slag inside the vessel during the oxygen blow. A third vessel was authorized in 1965 and went on line in 1966. A waste-heat recovery system for plant steam generation was added during 1965 and 1966.

As of 1964 Lackawanna's BOFs were charged with 70% hot metal and 30% scrap; massive hot-metal ladles and scrap-handling equipment cut charging time to three minutes. Oxygen was delivered through a 66' lance at 25,000 cfm. Limestone, burnt lime, ore pellets, and flurospar were periodically added to the vessel from overhead bins. The exact recipe of ingredients for each furnace charge was determined by a computer which also analyzed steel samples. However, computer control did not originally extend beyond the static system of charge calculation to include regulation of lance height or oxygen flow rate as a means of compensating for constantly changing conditions within the vessel.

BOF production rendered a portion of the plant's 35 open hearths redundant. The 14 obsolete units in #1 shop were shut down after 1964 for eventual demolition. On the other hand, open hearth production benefitted from injection of another European technology, vacuum degassing. This process was initially intended to cleanse molten steel of unwanted gases that could form flakes in the ingot, a potential source of weakness in forgings and other products. In 1957, Bethlehem Steel began purging hydrogen by pouring forging steel ingots through vacuum degassing chambers at its Bethlehem, Pennsylvania works using multi-stage steam jets to evacuate the pouring chamber containing the ingot mold.

Subsequent systems involved removal of oxygen as well as hydrogen; Erie Forge & Steel Corp. of Erie, Pennsylvania was the first to practice vacuum deoxidation

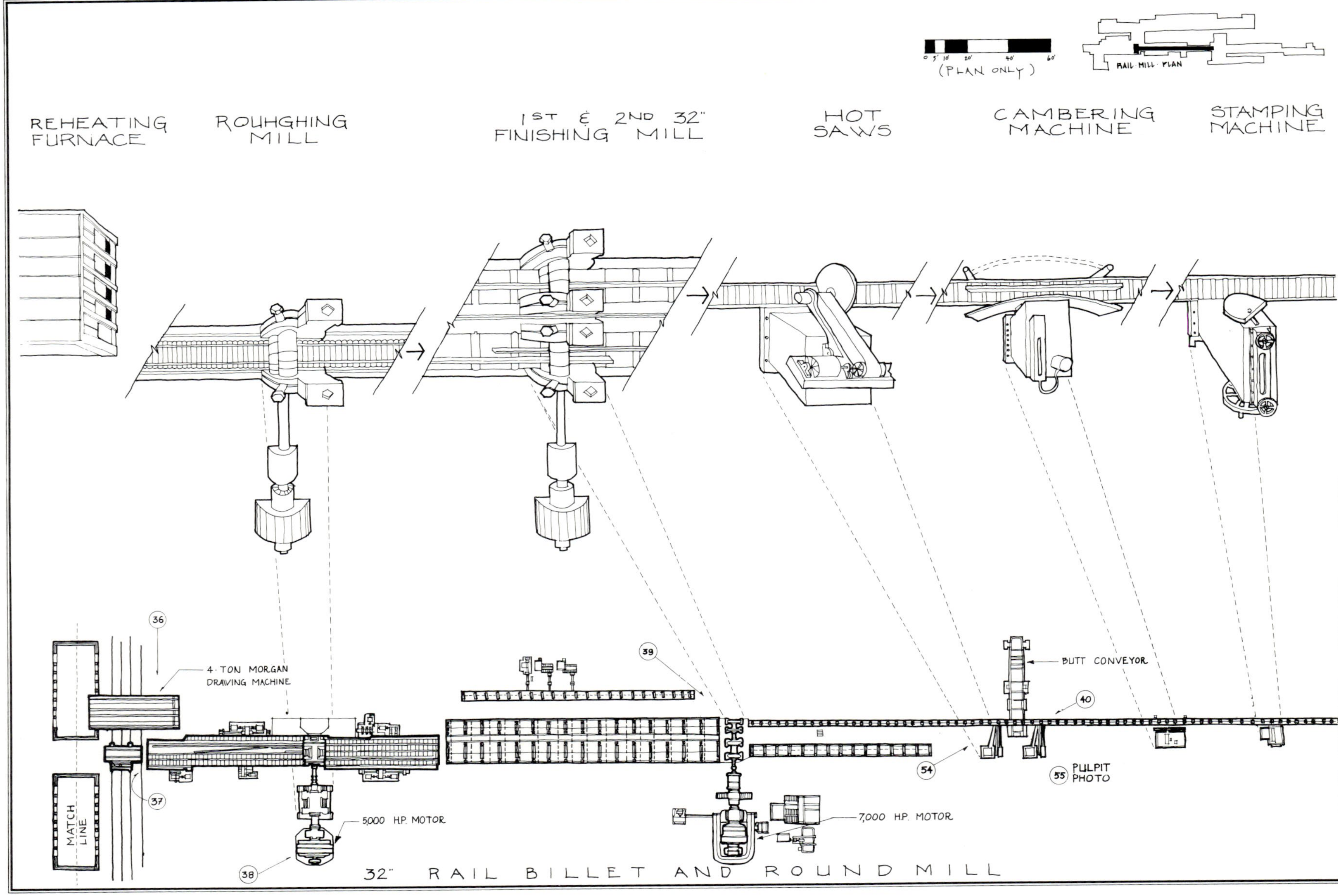

DELINEATED BY FRANCISCO HIDALGO-ROMERO · CAROLYN HONAN · ROB HAYES

BETHLEHEM · STEEL · PLANT BUFFALO · 1904
LACKAWANNA, NEW YORK

HISTORIC AMERICAN BUILDINGS SURVEY
HISTORIC AMERICAN ENGINEERING RECORD
NATIONAL PARK SERVICE
UNITED STATES DEPARTMENT OF THE INTERIOR

SHEET 2 of 4

HISTORIC AMERICAN ENGINEERING RECORD

on a commercial basis as of 1959. Vacuum treatments were also developed with the capability of adding alloys to heats as well as removing gases. In 1965 Bethlehem installed Dortmund-Hoerder ladle vacuum degassing apparatus at Lackawanna's #2 open-hearth shop; this German method was capable of degassing and alloying large tonnages. The difference in pressure between the ladle and the vacuum chamber extracted suspect gases from droplets of steel; carbon, manganese, silicon, and other alloying elements could be fed from storage into the degassing chamber by pushbutton controls. Vacuum degassing superseded the practice of deoxidizing heats by adding aluminum, silicon, or manganese before or after tapping, a method which was dependent on human judgment and therefore subject to some variation. Bethlehem's introduction of the D-H ladle process was intended to refit #2 open hearth as a specialty shop, making smaller heats of higher quality for such products as automobile gears and crankshafts while more common grades of tonnage steel were being shifted over to the BOFs.

Rolling Mills

Capital investments at the primary and finishing mills continued the trends established during the earlier years of the post-war boom. Lackawanna's temporary role as supplier of semi-finished steel to Burns Harbor stimulated some expansion of primary mill capacity. At the 45" x 90" slab mill, 9 more soaking pits were built during 1965-1966; the new pits featured individual electric cover drives rather than the less-efficient common cover drives used on the original 27 pits. However, the basic pattern of piecemeal improvements in production, conditioning, and inspection equipment continued.

Sporadic modernization concentrated on improving the quality of flat-rolled and bar products as the automobile industry tightened its product specifications. In 1963 Bethlehem announced a significant program for updating the strip mills to catch up with newer facilities at other steel plants. Each of the 7 finishing stands on the 79" hot strip mill was to be equipped with automatic gauge controls, including load-measuring devices, new screwdown motors, and roller bearing breaker blocks. New loopers regulated the tension on the strip between stands. The electronic generation of pressure sensing and speed controls was designed for more rapid and precise adjustment of the reductions in each stand that determined the final thickness of the strip. Comparable automatic gauge controls were also slated for the 75" tandem cold reduction mill.

During the late 1960s Bethlehem also began phasing in a new system of billet inspection and steel conditioning. Instead of detecting flaws by eyesight alone, inspectors began using black lights in conjunction with a machine that sprayed the billets with a solution containing phosphorescent magnetic particles and then concentrated the particles in defective areas via an electric charge. For removing those surface defects, Bethlehem subsequently planned to replace conventional chipping with four grinding machines of European design. Attempts to improve the quality of the bar production in the 10" and 12" mills included a new shot blast cleaning machine that removed surface scale or rust by abrasion instead of chemical action in acid pickling vats. The structural mills also saw limited improvements, including modernization of the 28" mill heating furnaces, new finishing equipment such as rotary straighteners and cold saws, and enlarged shipping facilities with additional double trolley cranes, transfer and roller tables, and runway extensions.

Lackawanna's limited share of Bethlehem's capital allocations during this period maintained overall finishing capacity at the level of the late 1950s but provided little scope for expansion. The rated capacity of the 48" and 28" structural mills increased a bit, but other facilities were written down including the 30"-21", 32", 14", 79", and bar mills. Piecemeal modernization left the mills with a mix of archaic and updated physical assets, a characteristic of the plant as a whole.

ROLLING MILLS
Plan of 32" mill.

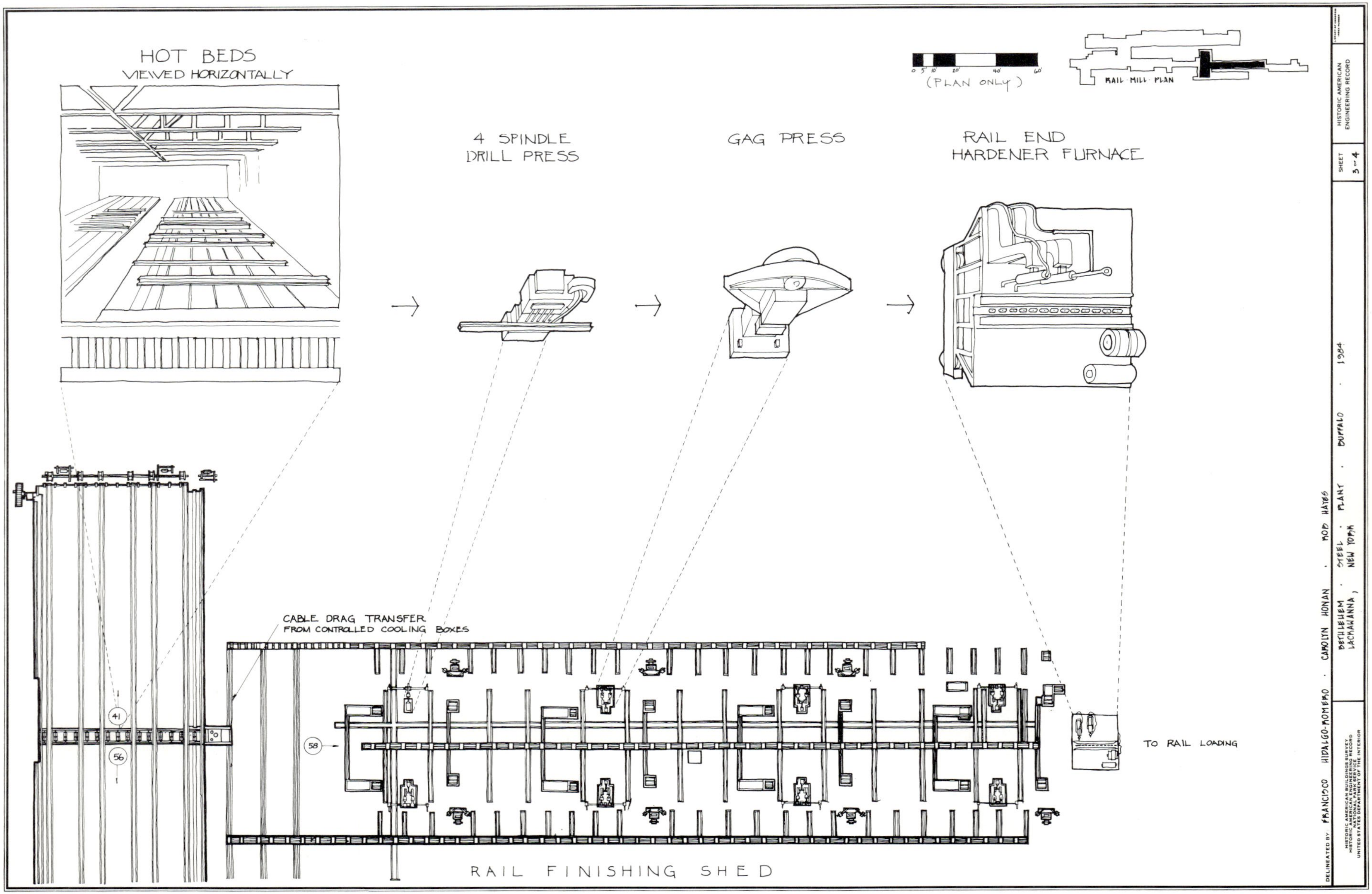
HOT BEDS
VIEWED HORIZONTALLY
4 SPINDLE
DRILL PRESS
GAG PRESS
RAIL END
HARDENER FURNACE
0 5' 10' 20' 40' 60'
(PLAN ONLY)
RAIL · MILL · PLAN
CABLE DRAG TRANSFER
FROM CONTROLLED COOLING BOXES
41
56
58
TO RAIL LOADING
RAIL FINISHING SHED
DELINEATED BY FRANCISCO HIDALGO-ROMERO · CAROLYN HONAN · ROD HAYES
BETHLEHEM · STEEL · PLANT · BUFFALO · 1984
LACKAWANNA, NEW YORK
HISTORIC AMERICAN BUILDINGS SURVEY
HISTORIC AMERICAN ENGINEERING RECORD
NATIONAL PARK SERVICE
UNITED STATES DEPARTMENT OF THE INTERIOR
SHEET
3 of 4
HISTORIC AMERICAN
ENGINEERING RECORD

6. WORK

Work and Technology

As the steel industry slumped during the early 1930s, an author named Horace B. Davis surveyed the prospects for social upheaval among its workers. He concluded that conditions had changed since the early years of the century when the modern industry was born out of fire and smoke to be baptized in the water of Wall Street stocks. The industry that had been expanding in 1910 was contracting in 1930. Turnover was decreasing as was the proportion of immigrants in the labor force. The percentage of older, more mature workers had gone up while unskilled laborers at the base of the job pyramid made up proportionally less of the work force than during the period 1890-1910.

While the composition of the labor force stabilized, so did the overall character of jobs within the steel plants. The authors of a study on the iron and steel industry under the 1933 National Industrial Recovery Act observed that, "Especially in the production [as distinguished from maintenance] departments . . . has mechanization done away with the validity of many old conceptions regarding degrees of skill that prevailed at the end of the War period." But technological change was never just a uniform pattern of unilinear evolution; jobs and skills changed at different rates in different departments. Still, Davis recognized that at many points in contemporary steel plants,

> The new jobs of operating new machines are different from the old jobs. They are more responsible than straight labor work, are frequently not under the direct control of a foreman, and are usually paid a little more than the common labor rate . . . They are neither skilled, nor semi-skilled, nor unskilled jobs; neither are they repair jobs. They are repetitive only in a broad sense; they do not make the worker a part of the machine; rather he runs the machine and controls it . . . They are something new, corresponding to the newness of the modern technique.

A plethora of major technical changes had restructured the context of work during the years when Lackawanna Steel built its new plant. By contrast, the period from the 1920s through the 1950s was an era in which a series of evolutionary improvements predominated. There were important exceptions, such as the construction of the continuous strip mills in the 1930s, where innovations transformed existing work processes at a single stroke. Even in the strip mills, however, some workers still faced an old danger from hot-mill cramps as rests between heats, common in hand mills, were eliminated. (Equally dangerous was the persistence of the physician's traditional remedy for this affliction, cocaine injections.) New practices, however, were more frequently designed to improve targeted phases of an operation rather than replacing the entire process.

Examples of this trend are abundant. At the coke batteries self-sealing oven doors superseded the older practice of manually filling the space between the door and the jamb with a special mixture of luting mud. Dolomite-spraying machines took over some of the chores that steelworker Charles Walker had performed by hand when relining the open hearths. Techniques for removing surface defects from semi-finished steel proliferated. Prior to the First World War chipping had consisted of knocking off large slivers with a hand hammer and cold chisel. After 1930 pneumatic hammers replaced the hand tools. Planing and milling machines likewise were introduced to remove surface imperfections from steel. Hand scarfing of cold steel with oxygen torches was attemped as early as 1919 but did not come into widespread use until the 1930s; mechanical hot scarfers, installed directly in the mill line so that torches could burn off blemishes in the steel's complexion while it was being rolled, were also a product of the inter-war period. Grinding machines represented another technique introduced into steel conditioning practice; in this instance a revolving abrasive cylinder was used instead of a single-point cutting tool or a torch for excising surface defects. In areas where semi-finished steel was stored, straddle carriers and bloom-turning machines mechanized the transportation and handling of materials.

ROLLING MILLS
Plan of 32" mill hot beds and rail finishing shed.

One by-product of gradual or narrowly-focused technological change was the preservation of work environments over several decades. In such situations the job structure and skills necessary for operating the facility might remain constant from generation to generation. Steelworkers of the 1950s and 1960s would still have to learn the same tricks and knacks of operating vintage machinery that their predecessors acquired during the 1920s and 1930s. This interaction between static technology and continuity in job skills could be found at older plants by mid-century. It is possible to observe a concrete example by walking through the historic 32" rail mill at Lackawanna, guided by Donald K. Landsittel who spent over 30 years working there.

Bethlehem's modernization program during the 1920s laid out the basic equipment that the 32" mill crews would use for nearly half a century when rolling rails. Rehabilitation of the mill included construction of reheating furnaces and elimination of both steam power and manual handling methods between passes at the first finishing stand. These changes altered the distribution of positions in the crew; by the 1950s, the number of men required to run the mill was also whittled down to an average of 35 men per turn. On the other hand, some of the jobs in the adjacent finishing shed remained largely untouched by major changes in technology. Straightening and drilling cold rails still required nearly as much heavy "bull work" as in 1903 when the first rails were produced at the plant.

In the 32" mill men eventually learned to work with equipment from another era. During the 1960s when crew members moved up through job progression, they had to cultivate some of the same operating know-how as steelworkers from the 1920s or even earlier. Since operating techniques remained relatively stable, filling a position in the mill also meant following directly in the footsteps of earlier generations.

ROLLING MILLS
Four-ton Morgan charging machine for loading sheared blooms from 44" mill into 32" mill reheating furnaces before rail rolling (1985).

The process of rolling rails began when the Operator of a mechanical charging machine deposited a sheared bloom from a transfer car into the reheating furnace. The Heater was responsible for bringing the bloom up to the proper temperature for rolling. For a number of years he regulated the furnace by adjusting chain and hand valves that controlled the mixture of air and fuel. The Heater was assisted by a Helper whose tasks included the hot job of "making bottom," or levelling the furnace floor. The Helper used pry bars and a water pipe to dislodge accumulated deposits that could otherwise cause uneven heating of the bloom. When the Heater judged that a bloom had reached rolling temperature, a drawing machine plucked the steel from the furnace and shuttled it over to the approach table where motorized feed rollers led to the 36" roughing stand. A mechanical bar turner was available to align the bloom with the 36" rolls if the drawing machine positioned it improperly. Operators for the charging and drawing machines were exposed to the high temperatures in the vicinity of the reheating furnaces; a Spotter relieved them periodically. A Furnace Recorder logged pertinent data as the blooms comprising orders from various customers moved through the reheating operation. Two Chasers originally kept the Heater in touch with the 44" blooming mill and subsequent operations in the 32"; they were replaced by a direct communication hookup later in the postwar era.

The Rougher had immediate charge of the 36" rolls where the process of shaping the reheated bloom into a rail began. He was also assisted by a Helper. The Rougher's responsibility included adjustment of the space between the rolls where the bloom was squeezed down to proper dimensions. It took a 15-20 lb. sledgehammer to activate the screw-down mechanism that changed the distance separating the top and bottom rolls. Maintaining the cross-sectional dimensions of the bloom within nominal tolerances required real athletic ability; the man swinging the sledgehammer stood on top of the roll housing and tried to maintain his balance while hot bars passed through the rolls underneath him. The Rougher Operator regulated the electrically-driven feed tables and rolls, plus a small run-out line, from his seat in the control pulpit near the 36" stand. He manipulated a total of seven controls, using his hands for the motors on the north and south mill tables and his feet for the main mill motor. Since the 36" rolls could accommodate as many as three bars at once, the Rougher Operator was constantly busy while the mill was rolling, even though he worked sitting down.

ROLLING MILLS
Four-ton Morgan drawing machine for extracting blooms from 32" mill reheat furnaces prior to rail rolling (1985).

Personnel stationed around the 32" first and second finishing stands included the Roller and the Assistant Roller. The Roller directed operations, supervising the crew, and acted as head technician. If the finisher rolls required adjustments during rolling, the Roller would measure test sections at his bench and then signal the Assistant Roller how much to compensate for the changes he detected in the section. By contrast with the rougher screwdown, the ring-and-pinion mechanism that adjusted the roll partings on the finishers did not require a sledgehammer massage. The Transfer Operator and the Finisher Operator worked in the pulpit near the 32" rolls. They coordinated the movement of hot steel bars on the mill tables and through the rolls in much the same ways as the Rougher Operator.

Other members of the crew working around the finishers included a Guide Setter and Guide Setter Helper who installed the equipment that controlled the entry and exit of the bar during individual passes between the rolls, an Oiler who basically acted as a handyman, and a Test Man who filled in as a helper when he wasn't carrying samples. Millwrights and electricians were involved with repairs or maintenance, but they did not come under the direct supervision of the Roller. Crane operators moved material around the mill, transporting rolls when the mill was being switched over to a different product, or helping clean up scrap from a "cobble" that had missed a pass, or even bringing the occasional test sample.

Once the rail had been formed in the roughing and finishing stands, the next step involved cutting it to standard lengths under the tilting hot saws. At this point, making the proper allowance for linear thermal contraction, or shrinkage of the rail as it cooled, became an important factor in maintaining product quality. A rail cut to 39' might shrink as much as 8" depending on the temperature at which it was finished; however, customers would accept a variation of only ±3/8" in length after cooling.

ROLLING MILLS
Approach table leading to 36" stand in 32" mill (1985).

It was the job of the Gauger to determine the exact point where the hot saw would make its cut. He judged the approximate temperature of each approaching rail by observing its color and then selected one of several paddles. The thickness of the paddles varied from 1¼" to ⅛". They were used to take up space between the end of the rail and a stop block. The Gauger chose whichever paddle would halt the rail under the saw so that the cut would include proper allowance for shrinkage. This job was one of the higher-paid positions in the 32" mill. Since sustained output of 70-80 rails per hour was not unusual, constant vigilance was required to detect the changes in finishing temperature that would affect paddle selection.

The Gauger stood right next to the roller line carrying the rails. He was thus exposed to the spectacular pyrotechnics of the hot saws which could send sparks shooting up to the full height of the 75' ceiling. Flying chips as big as nickels posed another safety hazard. The saws also made a lot of noise, but workers could tell from the sound what type of steel was being cut.

The Hot Saw Operator activated the cutting machinery from a pulpit, though the control area had not been enclosed during the Lackawanna Steel era. Other members of the crew around the saws included one Hot Saw Speller and Gauger who served as a reliefman and two Butt Pullers with tongs who removed pieces cut from the rail and deposited them into a conveyor for disposal. The Butt Pullers spelled each other every half hour. The Section Inspector acted as second-in-command to the Foreman of the finishing end.

The hot saws caused fireworks, but because their cutting capacity limited the pace of preceding operations they also created a production bottleneck. Don Landsittel kept daily notes during his career as a roller in the 32" mill; his records contain frequent references to problems at the saws. For example, during a turn on 25 February 1959:

> . . . Could not get mill going well & slow at saws. Running 1 & 5 in rougher [2 bars in the 36" rolls at once] — (slow) saws not taking steel — Kolodg and Bowen screaming cold steel — running slowest 1 & 5 pattern possible — still too fast for saws. Slowed mill down to single bar in rougher, single bar in finisher — 40-45 blooms per hour — rolled

ROLLING MILLS
General Electric 5000 hp reversing motor for driving 36" roughing rolls in the 32" mill. During the 1920s Bethlehem replaced steam engines with electric motors throughout the plant (1985).

last 2 heats this way — saws just keeping up at this speed. Slow roller line under saws, slow #1 saw action.

Once the rail had been cut to proper length, it then had to be cambered, or curved, in a slight arc. Cambering compensated for uneven cooling in different parts of the rail that could otherwise warp the finished product. The operation was performed on a machine consisting of a set of horizontal rolls that gripped the rail while a vertical roll on each side imparted curvature.

The job of running this machine shared some common characteristics with the work of the hot saw Gauger. The Cambering Machine Operator determined the exact amount of curvature for each rail by setting springs on the front vertical roll. Variations in finishing temperature meant keeping close watch on how the steel looked. Readjusting the springs was a tricky operation. The Operator also had to be prepared to replace worn bearings or make other minor repairs when the mill stopped rolling. Since the cambering machine was probably the oldest piece of equipment in the 32" mill, several generations of workers had to pick up similar knacks for adjusting it.

After cambering, the rail went through a stamping machine. Dies imprinted the heat number of the ingot from which the rail had been made and the cut letter that indicated its relative position at the top, middle, or bottom of a particular ingot. When the 32" was rolling rails, the Stamping Machine Operator was kept busy switching the dies for the heat numbers and cut letters, much like changing the movable type on an old printing press.

ROLLING MILLS
32" mill, first and second finishing stands with rolls removed. Entering from right, rails passed back and forth four times in first finisher (R) and once through second finisher (L) before being carried to hot saws in left background (1985).

The rail was then delivered to the hot beds where it began to cool, shrink, and straighten out. For many years Hot Bed Recorders were responsible for keeping tabs on the condition of the rails in this area. Those workers were eventually displaced by computerized record keeping, though the system's performance did not measure up to expectations. Turners, Swabbers, and Pulpit Operators shared the heat with the inspectors in this area where mechanized operations and manual work continued to coexist.

In Lackawanna Steel's original layout and for some years after Bethlehem's 1920s modifications, rails were transported from the hot beds directly to the finishing shed via a series of roller lines and transfer conveyors. During the 1930s the flow of production was modified by the introduction of controlled cooling, a process designed to forestall formation of shatter cracks that could eventually result in rail breakage due to undetected internal transverse fissures. Like cambering, controlled cooling compensated for uneven heat loss in different-sized parts of the rail; in this instance the problem of shatter cracks emanated from temperature differentials between the exterior and interior of the metal comprising the rail head. The rails remained on the hot beds until they no longer showed visibly red in a dim light, indicating that their surface temperature had dropped below 1000°. At that point the rails were transferred to the controlled cooling area where workers placed them in covered boxes using large magnets. Sealed in the controlled cooling boxes, the temperature of the rails dropped slowly to minimize the thermal stresses that could leave hidden physical defects. Having been properly cooled, the rails then proceeded into the finishing shed where they rejoined the original route of production.

At the finishing shed roller lines on either wall of the building delivered the rails onto inclined skids, but handling the rails after that point called for considerable physical exertion. The rails were moved manually from the skids to various machines; muscle power hefted finished rails onto a central roller line for delivery to the loading department. Visitors to this department over the years would have observed little more labor-saving machinery than had been present in 1903, though an air cylinder arrangement ((cannibalized from freight car brakes) was eventually installed to take some of the heavy work out of flipping the rails off the lateral roller lines and onto the skids.

The rail finishing department was known as the "mule shed" for good reason. Dragging the rails down the skids or turning a batch over for inspection was basically "bull work". However, the men who stayed on such jobs usually picked up the knack of where to position their feet and use leverage to best advantage. During inspection they also got a respite of about 10 minutes while the inspector walked over

ROLLING MILLS
32" mill hot saws cut rails to proper length as they moved from right to left coming from finishing stands (1985).

approximately a hundred rails that had been turned on the final bed; then it was time to repeat the process. Interlocking rails by turning them on the cooling beds prior to final shipment was another maneuver that called for human muscle to supplement machines.

Operations conducted in the finishing shed where job descriptions remained relatively constant over a long period of time included cold rail straightening on gag presses. Engineer John Fritz had substituted cold straightening presses for 60 lb. sledgehammers as early as 1857 when he revamped the Cambria iron works at Johnstown, Pennsylvania. However, straightening rails remained a heavy job, and those who did it were among the minority working 8-hour turns during the era of the 12-hour shift. The gag presses at Lackawanna required two men per machine. The job of the Straightener involved positioning each rail with its kinks under the movable top block of the press. The rails could weigh 1800 lbs. or more, and the Straightener walked them back and forth by hand. When the rail was in place he signalled to the Gagger by nodding his head. The Gagger, seated on a bench, then inserted a block of steel (the "gag") between the top block and the rail. Since the top block travelled a fixed distance on each downward stroke, gags of various sizes were used depending on the severity of the kink to be straightened. The force of the gag press bent the rail beyond its elastic limit and imparted a permanent set.

Straightening rails in this manner remained part of the work process in the 32" mill for several decades. The eventual introduction of welded rails made straightening even more critical, but also required borrowing a different method from the structural mills. More stringent specifications led to trials with rotary straightening during the 1970s. However, the new operation did not immediately surpass the performance of hand straightening on the gag presses.

ROLLING MILLS
32" mill hot beds where rails cooled prior to finishing (1985).

Rolling rails or other sections on a vintage mill like the 32" thus called for orchestrating the efforts of men with the capabilities of older machines. Various crew members, and particularly the Roller in overall charge, continued to exercise judgment and discretion in the course of their daily work. This aspect of the relationship between job skills and plant technology was particularly evident when setting up the mill for each different section and when making adjustments to maintain the desired section during rolling.

In order for the 32" mill to roll any of 17 different rail sections properly, the grooved upper and lower rolls had to be in correct horizontal and vertical alignment within their supporting housings. The Roller would check critical dimensions with a straight edge and calipers; feeler gauges, homemade from hacksaw blades, were used to measure the narrowest apertures between the top and bottoms rolls. Setting up the 32" mill also involved installing the proper guide equipment for each of the 10 passes necessary to form a finished rail. The guides controlled the horizontal and vertical path of the bar as it entered and left the rolls. Particular crew members were responsible for initially setting up the rolls and guides. The Roller's responsibility included checking their work as well as selecting the special combination of rolls and guides required for each section.

Proper set-up provided the foundation for accurate rolling, but holding the rail section within specified tolerances also called for keeping elements of the mill in dynamic equilibrium during rolling. As variables in the initial set-up changed, the Roller was expected to initiate the proper corrections for any deviations affecting the finished section. He integrated data from measurements on test sections with his own observations about the steel's appearance and behavior; through a synthesis of information and experience he determined which among several factors were affecting the finished rail and prescribed optimal compensatory adjustments for common section and surface defects. In effect, the Roller acted as a process computer under a green hard hat.

ROLLING MILLS
Rail finishing shed. Rails were moved manually from roller lines along either wall to inclined skids alongside gag presses and multiple-spindle drilling machines. The central roller line led to the loading department (1985).

The price of making satisfactory rails in the 32" mill was constant vigilence by crew members. As veteran Roller Don Landsittel put it, "The one thing that can be depended upon is that the rolls and section will not stay the way it is for any great length of time." Vigilence affected not only productivity and earnings but also safety. In an older mill the performance of the machinery was contingent on the quality of the crew who maintained human control over a plastic and fickle process.

Older facilities that continued running into the 1960s demanded men who could think like computers because those electronic marvels were never very far away. The latest wave of steel technology mixed larger or faster versions of staple processes with some new departures that could potentially affect both production methods and job skills. However, since some of the most recent innovations were not yet considered suitable for across-the-board applications, even high technology was destined to have an uneven impact on specific work situations in the short run.

The tentative introduction of process computers offers an example of this continuing trend from the earlier history of the steel industry. Rolling mills were among the first steel processes to incorporate computerization extensively. In 1962 McLouth Steel installed the first successful hot strip mill control computer capable of making adjustments while the steel was being rolled. By the mid-1960s process computers were in use on a number of hot strip mills, a large tonnage operation that was nearly continuous. At Bethlehem's new 80" mill in Burns Harbor computer control extended backward and forward from the actual rolling operation to include regulation of the firing rate on the slab reheating furnaces and the cooling sprays on the runout tables, all of which combined to affect finished product quality. Controlling the hot rolling process through computerization appealed to steelmakers because, in the words of one manager, "Rolling is extremely fast, and a computer has an understanding of relationships that an operator couldn't have."

However, computer technology did not prove immediately successful in plate mills. Although Republic Steel had introduced computer control on its Gadsden, Alabama plate mill as early as 1961, there were failures at other sites due to defects in sensing devices. X-ray gauges and load cells failed to monitor and transmit reliable data that were necessary for making rolling adjustments through an information loop rather than manual control.

Computerized steelmaking in basic oxygen furnaces was even more problematic in terms of the changing balance between operator control and analog or digital control. At issue was the old question of whether the skills of operators could be built into machines in order to improve corporate profitability, product quality, and managerial control over the production process. The advent of the BOF had rendered several decades of accumulated open hearth steelmaking experience irrelevant. In the words of a steel research executive,

> The open hearth is slow and gives time for human judgement. In the BOF, we go from hot metal to steel in 20 minutes, and we have no way to see the process as it occurs.

Some plants including Lackawanna, coped with the dynamics of the new process by installing control methods such as operator guide computers and static charge calculation systems that blended information processing with human judgment and reaction. Other firms chose technical options that converted BOF steelmaking into a servomechanical operation with a much greater degree of automatic control through computerization.

By contrast with the situation at Lackawanna's BOF shop, computer control systems of the second type were installed at Jones & Laughlin's Cleveland works and McLouth Steel in 1965. These new systems were designed to speed up the already-rapid steelmaking cycle in the BOF. During a heat in shops such as Lackawanna's, there were periods when the blow had to be interrupted for computer analysis of test samples obtained manually. As in open hearth steelmaking, these tests were necessary to measure two critical variables: temperature and carbon level. Since no reliable method then existed for continuous monitoring of temperature and carbon during a blow, the vessel had to be turned down, workers had to obtain samples from the molten bath, and then the heat had to be reblown or cooled depending on the test results. All these steps consumed time without actually making steel, although they were critical factors in determining the final quality of the heat.

The 1965 Jones & Laughlin and McLouth BOF control systems tried to reach the desired end-point temperature and carbon level automatically rather than through a series of successive approximations. Continuous analysis of exhaust gas content was used as a surrogate for continuous measurement of temperature and carbon. In response to the constantly changing conditions inside the vessel as revealed by the exhaust gas, computer control systems made mechanical adjustments that affected the progress of the chemical reactions required for making steel. The process could be accelerated or slowed down by increasing or decreasing the distance between the oxygen lance and the bath as well as by regulating the oxygen flow rate. This servomechanical linkage between exhaust gas monitoring and mechanical responses was intended to permit more rapid blowing without danger of slopping. Engineers felt that since the computer reacted more quickly than an operator, the control system would save some of the minutes presently lost in making tests and corrections.

The closed-loop, automatic control system was by no means universally accepted in the mid-1960s despite its potential for deskilling operators. Questions were raised about the feasibility of devising a mathematical model for reactions occurring during BOF heats and about the reliability of inferring temperature and carbon levels from exhaust gas content. Since Bethlehem was one of the skeptics, the original computer system at Lackawanna's BOF shop left more responsibility in the hands of human operators.

The redefinition of BOF operators' responsibilities depended to a large extent on the exact capabilities of whatever computer system corporations chose to install. However, even the Jones & Laughlin system, developed in conjunction with Westinghouse, was purposely introduced as an open loop where operators responded to computerized analysis by initiating some control responses themselves. J & L apparently felt that preserving elements of manual control served as an important preliminary step; it built up operator confidence in the system before the conversion to a fully-automatic closed loop.

Computerization of steel plant operations remained in its infancy during the mid-1960s when Bethlehem's Burns Harbor mills were just going on line, and Lackawanna's older plant was straining to capacity. Previous eras of significant change in plant technology and work processes had also featured a similar mixture of the old with the new. Nonetheless, the overall trend had been toward increasing profit, productivity, and corporate control over production through intensive mechanization.

However, around 1970 the piecemeal upgrading of existing techniques had reached a plateau. Labor cost reductions were less than the expense of new capital investments. Rising capital costs offset the savings from modernization even at new plants such as Burns Harbor where Bethlehem had sunk a billion dollars into the largest privately-financed construction project of the 1960s. The ratio of capital input to unit value of output in the steel industry had doubled during 1947-1967. One consequence of steel's rising capital-output ratio was the relative decline in the percentage return on net worth which corporations such as Bethlehem began lamenting during the 1970s.

Conversely, Bethlehem and other large integrated companies were slow to adopt some of the recent innovations capable of improving plant technology by dramatic leaps rather than just incremental upgrading. For example, Bethlehem's decision not to install continuous casting at Burns Harbor in the 1960s was accepted as conventional wisdom by much of the steel industry; a Bethlehem vice-president felt that continuous casting was a decade away from being able to deliver the quality and output needed for the new plant's flat-rolled product mix. When steel corporations began having difficulty recouping their profits, they attempted to economize by cutting costs without implementing major changes in existing plants such as Lackawanna. This strategy was destined to cause a confrontation between labor and management over such issues as productivity and work rules. The conflict would be fought out within the system of industrial relations that had emerged during the decades when technology remained comparatively static.

Work and Authority

Although the transformation of job skills slowed down as the pace of technological change stabilized through the 1950s, labor relations changed dramatically. The profound economic crisis of the 1930s provoked a seismic shift in the balance of power among corporations, workers, and government. During the Lackawanna Steel era, company and plant management had monopolized legal authority over decisions affecting the jobs of steelworkers, though some policies were occasionally modified through public pressure or legislation. When Bethelehem Steel took over, it continued the same mixture of paternalism and repression.

The Great Depression fractured the era of comparative labor stability and rekindled a rank-and-file militance that permeated the steel industry. Discontent eventually coalesced in a drive to organize all steelworkers into an industrial union whose jurisdiction would cover an entire plant — the Steel Workers Organizing Committee or SWOC. This form of organization differed from the 1918-1919 campaign to recruit workers for existing craft unions, but the agenda still included the same basic issues that had been left unresolved by the 1919 strike. Federal support of employees' legal rights to form unions plus pressures to maintain defense production were both instrumental in eventually establishing collective bargaining at Lackawanna and Bethlehem plants during the 1940s.

However, the powers of the law and the state also frustrated the aspirations of organized workers for more say on the job. During the postwar era, management, labor, and government developed contractual provisions and bureaucratic procedures for channeling discontent into authorized outlets. This compromise arrangement made an important difference in the process of resolving disputes between workers and supervisors in the plant. Job classifications and wage scales were standardized for the first time. Some foremen had been guilty of favoritism and abuse of authority during the Depression years; those practices were curbed as well. The remote corporation that had seemed so inscrutable to novelist Thomas Bell now at least acquired a human face, even if it only appeared across a table during negotiations.

On the other hand, it is equally true that corporations' legal authority over daily work and production remained largely intact within the new system of industrial relations. Specific contract language affirmed management's right to direct and discipline the work force. A complex 5-step grievance procedure culminating in impartial arbitration provided steelworkers with a legal instrument for challenging unauthorized work rule changes or other contract violations. Constructed as a defense against the petty tyranny common during the open-shop era, the new apparatus of industrial jurisprudence also enabled corporations to police wildcat strikes, slowdowns, and other acts of rank-and-file militance through the collective bargaining agreement itself. Grievances still boiled over into work stoppages at Lackawanna, but direct action now violated union contract provisions as well as management prerogatives. The elaborate architecture of rationalized labor relations supported and confined workers' protests against management decisions.

When Bethlehem Steel began rebuilding the Lackawanna plant during the 1920s, the local Chamber of Commerce journal proclaimed that the new program would reconstruct "morale as well as machinery." Bethlehem put up new welfare buildings where workers might change clothes or wash up at the end of a turn. The corporation initiated a contributory Relief Plan that offered death and disability benefits in cases not covered by workers' compensation laws. Employees with 25 years of service were eligible for pensions up to $100 per month; a pensioner could take another job ". . . unless he is advised his proposed employment is deemed prejudicial to the interests of his former employers." Through a subsidiary, the steel company developed a new residential area called Bethlehem Park, consisting of 300 houses on 63 acres south of Smokes Creek and east of the Hamburg Turnpike; families from different provinces in Italy eventually purchased many of the homes in the new settlement. Meanwhile, the tenements in the Old Village built by Lackawanna Steel were at first used for housing black workers and then scheduled for demolition as of 1930.

Bethlehem also installed an Employee Representation Plan (ERP), derived from the corporation's experiences with labor organizers and government agencies during the First World War. Under the rubric of introducing "industrial democracy" into plant life, the ERP provided a forum where elected representatives presented departmental grievances for management's consideration. However, the program was essentially a superficial reform that left the legal disparity between corporate authority and workers' rights untouched.

Through these welfare programs and labor policies Bethlehem sought to mold a more stable and tractable work force. Local business students visiting the plant in 1924 discovered that, "Like every other human workshop, the Company is looking toward its dollars and cents profits rather than toward charity or benevolence." Investments in new washrooms or housing might cut costs associated with high rates of labor turnover and save the company the expense of constantly breaking in new men. Improved working and living conditions might likewise innoculate Lackawanna's employees against any recurrence of the union sentiment that had broken out in 1919. The ERP's "orderly and expeditious procedure for the prevention and adjustment of any future differences" allowed the corporation to keep a firm grip on the reins of power while promising cooperative workers a smoother ride.

None of these policies guaranteed any real job security when production and employment plummeted during the Depression. By December, 1931 the company admitted that its effort "to spread available work over the payroll as widely as possible" had not provided a living wage for Lackawanna's 7000 workers. It offered to furnish relief work for all those earning less than $12 per week, though only 250 men were being aided by the program a month later.

Individual job security was eroded still further when *de facto* control over employment reverted to the shop foremen. With work scarce, foremen reasserted their influence over hiring, layoffs, and promotions, bypassing the central employment office in the process. Previous corporate gestures toward more enlightened — or, at least, more efficient — personnel practices became irrelevant when supervisors resumed dispensing favors and discipline according to personal whim. Under these circumstances getting and keeping a job meant currying favor while surrendering dignity in a multitude of subtle ways.

Neither relief schemes promulgated from Bethlehem's home office nor the ERP addressed the basic problem of job security. The ERP at Bethlehem plants gained some credibility from its relative longevity compared with other company unions that were conjured up merely to circumvent New Deal labor legislation. But the program had no contractual agreements enforceable through strikes, no source of funds independent of the company, no staff of expert advisers to muster evidence defending workers' positions on complex issues, no framework for considering general issues that went beyond departmental grievances. Given these defects, it is not surprising that resentment against the regressive, arbitrary, and discriminatory attitudes of foremen and other supervisors eventually outweighed support for the company union.

The majority of Lackawanna's steelworkers tried to gain some control over their jobs by demanding labor policies that were more rational than the company's current practices. Measures such as determination of work assignments according to seniority, job classifications with specific pay grades and responsibilities, and grievance procedures to deal systematically with violations of contractual provisions would provide a line of defense against the inconsistent authoritarianism that had resurfaced inside the steel plant during the Depression. Advocates of industrial unionism pressed for these reforms during the late 1930s and early 1940s.

Organizing at Lackawanna took place within a larger political and economic context. The National Labor Relations Act (Wagner Act) affirmed the legal rights of employees to bargain collectively through representatives of their own choosing. Bethlehem's ERPs were held to be violations of the Wagner Act and were eventually disestablished. As business conditions gradually improved, federal priorities shifted from coping with a depressed economy toward preparing for confrontation with fascism overseas. The Steel Workers Organizing Committee drew on a cadre of organizers and support staff from the Congress of Industrial Organizations (CIO) which had broken off from the American Federation of Labor to become a dynamic rival. The heavy mass-production industries were then the most advanced sector of American capitalism, and the labor movement had to adapt to the changes in the basic structure of the economy that Bethlehem Steel Corporation and the Lackawanna plant represented.

The CIO set up a regional office for its steelworkers organizing campaign in early August, 1936. The initial open meeting in Lackawanna was scheduled for 18 October 1936 at Friendship House, 195 Ridge Road. Slated speakers included the Rev Francis J. Hunt, pastor of Our Mother of Good Counsel Church in Blasdell, on "Social Aspects of Labor," and the Rev. Lewis W. Holley, Buffalo Council of the National Negro Congress, on "The Steel Drive and the Negro People."

Nationally, SWOC got off to a fast start. In March, 1937 U.S. Steel, the citadel of the open shop during Judge Gary's reign, agreed to recognize SWOC as a bargaining agent in the plants of its Carnegie-Illinois subsidiary and subsequently in its other holdings as well. However, Bethlehem and a group of other large corporations known collectively as "Little Steel" resisted the notion of diluting managerial authority within the workplace. The Little Steel drive bogged down amid periodic outbursts of violence in plants across the Northeast and Midwest. As an early historian of the industrial union movement in steel recorded, by 1939

> . . . the organizing campaign against Bethlehem was settling down to a long-run educational program in which the primary emphasis was on the development of leaders from the bottom up, and detailed training in the elementary techniques of collective bargaining.

This emphasis on grass-roots education bore greatest fruit at Lackawanna where support for SWOC surpassed other Bethlehem plants.

One result of growing SWOC strength was the evolution of an informal grievance adjustment procedure involving representatives of labor and Lackawanna plant management, even though the union as yet had no legal status as a bargaining agent. Similar developments had occurred at other Little Steel plants. There was a complete shop steward network in place at the Indiana Harbor plant of Inland Steel by the fall of 1939. Plantwide grievance machinery was also active in the Mahoning Valley plants of Youngstown Sheet & Tube by January, 1940. In addition to demands for wage adjustments which were often filed on behalf of particular job classifications, common subjects of dispute at Lackawanna in 1940 and early 1941 included issues related to safety, discrimination against union members, and work rule violations, as well as the effort to establish seniority as a basis for rehiring or promotion.

ROLLING MILLS
Crew of the #5 48" universal and sheared-plate mill (1929).

Safety grievances included requests to alter work routines and manpower requirements on certain jobs. When four men on the battery crew were overcome by excessive heat at the coke ovens, SWOC suggested that relief men should be available in the future. Another grievance called for doubling the crews on steam locomotives serving the blast furnaces to include an engineer, a fireman, and two switchmen. Issues raised concerning personal safety included questioning management's wisdom in sending a 63-year-old oiler up on top of the giant ore reloading bridges at night. Blast furnace larry car operators who worked in a relatively warm environment objected to being ordered to climb up the furnaces in mid-winter without provision for proper protective clothing, and men at the splice bar mill likewise complained about working out in the rain without coats or hats. In an inherently dangerous occupation there were always legitimate concerns about unsafe equipment, as indicated by the following grievance from 27 November 1940 about the milling machines and drill presses in the rail finishing shed:

> These machines are antiquated and out of condition. The life and limb of the men working about them are in danger. Three men have been injured in the recent past. Even the bosses have said the machines are not safe. And the foremen have promised that the machines would be replaced by better ones. But nothing has been done about it, and men continue to be hurt. The Company must correct this condition at once. All ready the men in the Rail Mill are inflamed that the Company cares nothing for the safety of the men. The Company will not be permitted to use dangerous machines indefinitely at the needless sacrifice of the life and limb of the men. The men request the Company to rectify this condition at once.

ROLLING MILLS
Catchers loop steel by hand through finishing stands of the #10 8" bar mill (1929).

Grievances protesting discrimination against SWOC activists usually involved demotions or questionable work assignments by foremen. A pro-union man who had been operating an iron mule at the coke ovens was downgraded to a laborer's job. SWOC members in the blast furnace crews who reported off for Christmas were docked a day's pay, but men less vocal in their union sympathies were not penalized for the same action. Management blocked the promotion of SWOC members at #3 open hearth by hiring new men as second helpers in #2 open hearth and then transferring the displaced second helpers from #2 shop over to #3 shop. As a result, third helpers in #3 shop were deprived of the opportunity to become second helpers. The finishing end foreman in the 32" rail mill tried intimidating the men under his authority by arbitrarily sending some workers home before they had completed an 8-hour turn, thus cutting their earnings. In the electrical shop a layoff of 27 men involved discrimination against active SWOC members.

Grievances over disregard of seniority represented labor's attempt to reform personnel practices by introducing consistent standards that would counteract the unpredictability and favoritism left over from the Depression era. One of the earliest grievances filed at Lackawanna during the period prior to formal union recognition involved a man who had begun working at the storage beds in 1928. He had been laid off on several occasions, the last time in 1937. Nevertheless, SWOC pressed for his right to be recalled since men with less service had been hired into the department. A worker in the rail mill believed that length of service should have entitled him to certain proprietary rights concerning his usual job assignment:

> This past week I was taken away from my job of Transfer Operator and put at common labor and another man has been put in my place of transfer operator. During the hot weather I had the hot job of transfer operator — now that cool weather has come they give me work as common labor and take my old job away from me and give it to someone else. Also they have shifted me from the B-turn to the C-turn, and the man now working on the B-turn at my old job is a man with only one year service at that kind of job while I had 12 years service in the Hot Bed.

ROLLING MILLS
Turning blooms manually. This task was mechanized by the early 1950s (1929).

During the latter part of 1940 simmering labor tensions flared up into partial work stoppages as the pace of organizing at Lackawanna accelerated. In August 500 men at the coke ovens staged a 1½-hour protest over Bethlehem's slow response to SWOC demands for wage increases. In mid-September 2000 men from the blast furnaces and ore docks launched a walkout after their superintendent refused to meet with a grievance committee about two union members who had been demoted; 2000 more men, including 150 from the coke ovens, soon joined in. When the open hearth departments threatened to halt production as well, a mediator from the U.S. Labor Department interceded. As national negotiations between SWOC and Bethlehem over minimum pay rates on government work and other issues dragged on into the fall, Lackawanna union leaders began worrying that their members would take the initiative in calling an unauthorized strike. For example, on 16 October union men in the 32" rail mill demonstrated against the dismissal of a laborer in the coke ovens who was also a member of SWOC.

However, the decisive confrontation at Lackawanna did not erupt until February, 1941. As had been the case during the sitdown strikes in the automobile industry, the technology that had made integrated industrial operations very efficient also proved highly vulnerable to work stoppages in key areas. Beginning on 7 February workers at the coke ovens engaged in a series of slowdowns that cut both fuel production for the blast furnaces and gas supplies for other departments. Plant management suspended hundreds of men from the coke ovens, ore docks, and blast furnaces; the suspensions then became an issue along with the original demands concerning wages and adjustment of grievances outside the moribund ERP. The crisis atmosphere increased SWOC recruitment substantially.

A 38-hour strike at Lackawanna ensued during 26-28 February. The strike was successful despite lacking unanimous supprt. For example, at the relatively new strip mill department there was violence on the picket line when non-strikers tried to enter the plant in automobiles, and one union observer recorded the names of 113

ROLLING MILLS
An army of chippers conditions the surface of shell steel billets during World War II defense production.

C. N.
130476

men who continued working in the 79" mill. Outside the gates, the attitude of local political authorities toward the strife turned out to be just the opposite of the 1919 alignment. In 1941 Lackawanna police and the county sheriff did not exert themselves unduly when Bethlehem called for assistance in controlling pickets, but Buffalo police ran interference for strikebreakers.

However, the strategy and settlement of the strike depended most heavily upon intervention by the U.S. government. The Office of Production Management, whose major interest lay in maintaining the manufacture of defense materials, obtained Bethlehem's consent for an election which determined that an industrial union would represent Lackawanna's 14,000 workers.

Once the basic questions of union recognition and collective bargaining had been decided, pressures from federal agencies began shaping the rituals of conflict that would endure during the postwar era. Wildcat strikes and other forms of direct action were discouraged by a coalition of federal administrators, leaders of the United Steelworkers of America (as SWOC was renamed in 1942), and corporate officials. Workers were told to resolve their protests through the grievance machinery or trust federal agents such as the War Labor Board (WLB) to decide matters. Variations on this scenario were played out frequently at Lackwanna during World War II.

In July, 1942 900 workers struck at #2 and #3 open hearths. They were at loggerheads with plant management over accumulated grievances, the posting of seniority and job lists, and the thorny issue of requiring 100% union membership within the bargaining unit. Strikers resumed work at the request of the WLB which later unclogged the grievance machinery and supported maintenance of union membership.

ROLLING MILLS
Structural steel shipping yard called "The Slaughterhouse" because of the high accident rate. Looking north from runway #8 to covered runways #3 and #4 (1929).

In January, 1943 125 electrical maintenance workers went out for 6 days when three men filed a grievance over how their pay for work on an ore bridge had been computed. Plant management turned down a proposal from a War Production Board consultant on the grounds that the union had not conformed to the grievance procedure. Workers from the machine shop, rigging department, and narrow gauge railway staged a sympathy strike that led to eventual settlement on the basis of the WPB's original proposal.

In June, 1943 185 men struck when management disciplined a shop steward in the 14" mill. Defending the steward's position, union reprsentatives refused to terminate the walkout so a federal conciliator was brought in.

In July, 1943 production at the 10" mill halted for 10 days after 8 men on the first shift refused to work on the hot bed at prevailing rates. Steelworkers officials subsequently announced that the strikers would return to work while the issue was pursued through the grievance procedure.

In March, 1944 another dispute was resolved in a similar fashion when 150 men agreed to return to the structural shipping department. At the urging of union officials the strikers agreed to file grievances about the unsafe conditions that had precipitated their brief walkout.

In June, 1944 140 electrical maintenance workers reiterated their two-year-old complaint about wage rates on repair jobs. A federal conciliator entered the fray once again.

In May, 1945 a major dispute mushroomed out of a company decision to cut two positions off the 8-man crews that unloaded raw materials for the open hearths. Within three days 5000 men — nearly half the plant's work force — had gone out. Local USW officials and international union representatives had to walk a fine line in dealing with the spontaneous and unauthorized walkout. On one hand, the leadership felt obligated to oppose arbitrary job elimination by the company; on the other hand, they also faced the necessity of convincing the rank-and-file to go back to work under a directive from the War Labor Board. At first 3000 strikers voted to continue their walkout, but reversed the decision by a 602-308 tally two days later after speeches from Steelworkers officials, the War Labor Board's director of disputes, and military personnel; the WLB agreed to adjudicate the original dispute.

Steelworker holding broken safety glasses (1929).

No.9 MILL ROLLS 'SAFETY RECORD
ACCIDENTS MINOR AND LOST TIME

13
Down with Capital
Vote
Socialism
NCR
PRR

Some of the dynamics of wartime collective bargaining carried over into postwar industrial relations at Lackawanna and throughout the steel industry. As the result of a 1944 WLB ruling, the steel companies and the union standardized job descriptions and basic wage-rate classifications (but not incentive programs regarding pay for extra output). This formalization of the job progression in the various departments further diminished the discretionary authority that foremen had wielded and abused during the Depression.

The shifting legal atmosphere after the war altered foremen's stature in another way. While steelworkers were joining their union, supervisory workers at Lackawanna also tried to form a collective bargaining unit comparable to the fledgling Foreman's Association of America begun at Ford during this period. A union of foremen threatened Bethlehem's efforts to reassert corporate control over the newborn system of industrial relations. The company may have feared a *rapprochement* between the USW and the foremen despite a long-standing tradition of hostility between workers and supervisors. In any event, Bethlehem went to court, and as of 1947 succeeded in invalidating a ruling by the New York State labor relations board that had certified supervisory workers as an appropriate collective bargaining unit. That same year, the Labor-Management Relations Act (Taft-Hartley Act) expressly excluded supervisors' unions from the provisions and protections of the Wagner Act.

Bethlehem's opposition to the Lackawanna foremen's union was just part of a multi-pronged offensive designed to insure that collective bargaining did not jeopardize the corporation's ability to discipline and control its work force. It was evident that steelworkers at the plant had learned the lessons of solidarity and cultivated a taste for the power it could confer. As noted above, during the war whole departments might turn out in protest against management actions that directly affected only a few men, and this tradition carried over into the postwar decades. Even at the strip mills, where support for the 1941 strike had been spotty, workers adopted the new rules of the game. In September, 1945 2000 men from the strip mills conducted an unauthorized walkout in support of 63 picklers involved in a wage dispute; 1200 men staged another quickie in October, 1948 that shut down the 79" mill when employees in the finishing end protested being assigned additional duties.

Safety meeting in #9 10" bar mill under Bethlehem's Employee Representation Plan (1930).

Over the quarter-century following World War II, labor-management disputes at Lackawanna that flared up into walkouts were eventually routed into the grievance machinery to comply with contractual provisions. Workers still halted production without union authorization for relatively short periods, but the threat of local strikes functioned mainly as a pressure tactic for introducing issues into the negotiating process rather than forcing an immediate settlement. Corporate and plant management retained the core of its responsibility and authority over investment decisions and daily operations; the union preserved the gains it had made by providing workers with some legal basis for challenging certain management decisions. Two examples can illustrate this routinization of labor militance during the rebalancing of authority relationships at Lackawanna.

The first episode began on 27 October 1950 when 750 men left their jobs in the blast furnace department. Their protest involved a company plan to hire new employees off the street for the new sintering plant rather than opening the higher-paying positions there to qualified workers from the blast furnaces. The next day 350 strikers rejected a union plea that they return to work pending a settlement through the grievance procedure; pursuing the case to arbitration could take as long as 7 weeks while the sintering plant could still begin operating with new hires during the interim. However, on the 29th the strikers did agree to return while a union-management conference hashed out the grievance over hiring rights.

Within 3 days turmoil enveloped the entire plant because of a situation that grew out of the original grievance. Bethlehem had fired Vincent Copeland, a grievanceman in the blast furnace department and refused to reinstate him despite appeals from the top committee of local steelworkers' officials and representatives of the international union. The word was out that Copeland was a marked man because of his role in orchestrating the earlier walkout and his general reputation as an agitator. Copeland sprang from a line of respectable dissenters and social gadflies, but he had journeyed well beyond the pale of conventional politics in joining

Switchman on South Buffalo Railroad (1931).

the Socialist Workers Party inspired by Leon Trotsky. Some of the steelworkers at Lackawanna had refrained from supporting the sintering plant walkout because of Copeland's involvement and political views. But Copeland's discharge violated an implicit understanding that management would not penalize the men who had participated in the wildcat strike. Members of USW Local 2601 representing the blast furnace and coke oven workers again left their jobs in protest.

Representatives of the international USW headquarters called a meeting for the evening of 2 November to vote on ending the wildcat. Over 2000 men jammed Legion Memorial Hall and another 1500 waited outside. Union solidarity eventually prevailed over political discord as the steelworkers voted to continue striking in support of Copeland. Once that point had been made, the new routines for settling disputes took over. Arbitration, not direct action, would decide Copeland's fate, and the following night workers voted 3278-924 to return to their jobs.

Another episode during 1953 exposed the latent tensions inherent in the new mode of industrial relations even more clearly. The controversy this time revolved around management's effort to change operating practices which would alter the job progressions standardized during the postwar reclassification program. However, the conflict quickly expanded into a prolonged confrontation that eventually pitted Lackawanna workers against the top leadership of their international union as well as against management.

The disputed change involved job assignments at the coke oven batteries. Management's plan would not diminish the overall number of men in the battery crews but did affect specific job assignments of 12 individuals. Previously these men had operated coal larries, pusher cars, and door machines as part of six-man crews; under the proposed division of labor they would revert to working as lidmen or door cleaners on a smaller number of nine-man teams.

Machine operators affected by the new arrangement interpreted the compulsory reassignments as demotions. Lidmen earned an hourly rate of $1.68½; car operators were paid $2.01½. Lidmen also worked under worse conditions with heavier exposure to gas and soot since they were stationed on top of the batteries to remove covers from the holes where coal was charged into the ovens. The ambient temperature here often reached 160°, and lidmen had to wear special footwear when treading across the ovens. Coke oven workers also claimed that the new job policy increased work loads. Formerly, the crews had each tended a total of 32 ovens; now they would be responsible for 48, which management held was equivalent to practices in other plants.

However, the basic objection to the change boiled down to its effects on job advancement. By removing some of the rungs in the battery crews' promotion ladder, the new policy froze more workers in the dirty jobs while giving fewer men the opportunity of moving up to the machine operators' positions.

On 31 August a wildcat strike began in the coke ovens when 300 men on the 7 a.m. shift refused to start work until these issues, affecting a dozen of their fellow workers, were resolved. The conflict spread to include 250 electricians in Local 2602 when three men were sent home after refusing to perform work for supervisors operating the strikebound coke oven department. Blast furnace workers also joined in. By 10 September production had ceased throughout the plant.

USW district director Joseph P. Molony asked the coke oven workers to return to work and make their point through the grievance machinery. Edmund Martin, general manager of the Lackawanna Plant, took management's usual stance. He announced that any negotiations to resolve the dispute through contractual channels would not begin until the wildcat strike was over. The Federal Mediation and Conciliation Service stayed on the sidelines.

As in the Copeland case, the situation became more complicated when management disciplined leaders of the walkout. Three coke oven shop stewards, Willie Hicks, Arthur Hicks, and William Brown, were suspended along with a fourth steelworker, Lee Harrison, for fomenting a strike in violation of contractual provisions. It was the stewards, not USW staff representatives, who publicly articulated the reasons behind the wildcat. USW leaders, committed to following the new rationalized bargaining practices, now faced the double task of urging the men to call off the strike while trying to persuade management that the suspensions posed the biggest obstacles to an orderly settlement.

On 3 September 1000 coke oven strikers rejected a USW staff representative's appeal to return to work pending a settlement through contractual channels. The strikers voted to disregard a similar appeal from Molony on 8 September. After conferring with USW president David J. McDonald, Molony planned another appeal at the regular bi-monthly meeting of Local 2601, which included workers from the blast furnaces, ore docks, and open hearths as well as the coke plant; he had wanted to tell the entire USW membership at Lackawanna that "a small group should not dictate" the actions of 17,000 members. However, on 14 September 3000 members of Local 2601 upheld the principle of workplace solidarity by voting unanimously to reject the recommendations offered by their international union. They were supported by several unions and the Greater Buffalo CIO.

With a major confrontation shaping up, McDonald personally intervened on 17 September and ordered the strikers to resume work. He condemned the wildcat as a violation of the union contract and followed plant manager Martin by maintaining that the grievance and arbitration structure provided adequate recourse for the men. International union officers would see that employees discontinued slowdowns, performed work that was properly assigned by supervisors, complied with contract provisions, and processed grievances through established procedures. A USW staff expert on industrial engineering would now assist Lackawanna workers in devising a realistic response to management's proposed changes.

These two episodes reveal some of the tensions and contradictions built into the postwar structure of industrial relations. Authority over the workplace was parcelled out unequally among management, government, and organized labor, who hammered out standard procedures for negotiating contracts and resolving grievances while trying to control both unilateral corporate power and the spontaneous protests of workers. Engineered by arbitrators, the formal and rational edifice of collective bargaining became as characteristic of the steel industry as monumental blast furnaces. However, the system could not bear the full brunt of persisting management-labor conflicts over workplace issues. While nationwide steel strikes were averted under the Kennedy and Johnson administrations, brush fires over local issues nevertheless kept breaking out at Lackawanna. Prior to the expiration of the 1968 national contract, plantwide negotiations at Lackawanna settled 247 outstanding issues, but left 362 unresolved. The grievance machinery groaned under the strain of handling such a load. Therefore, as the decade and an era drew to a close, both plant management and local union officials operated within a set of industrial relations that combined bureaucracy with unresolved confrontation.

FOREIGN STEEL
"STEALS" JOBS
WRITE CONGRESS
NOW
DO NOT ENTER
5480-KX

SHUTDOWN 1970-1983

7. THE CORPORATION

In June, 1970, Bethlehem Steel Chairman Edmund F. Martin toured the Lackawanna plant where he had been General Manager. He declared that Lackawanna was "the best damned steel plant in the country" and promised to keep it that way. Six months later Martin retired, leaving the plant and the company in new hands. Different company officers took the reins of power at the exact moment when national and international events began to occur which would affect all future corporate plans and lead to Lackawanna's demise 13 years later.

By 1970 much of the post-war economic boom was over. Most of the reconstruction was completed, and industries, both national and foreign, were operating at full capacity. The demand for steel to support new growth simply faded away. Simultaneously, new overseas industries, built and financed in part by American efforts, were now strong enough to challenge U.S. industrial dominance, often surpassing domestic manufacturing with more modern and technologically efficient production. Trying to keep profit rates high, American industries began raising prices, and the combination of inflation with declining productivity from shopworn mills produced the infamous "stagflation" of the 1970s.

Competition for crucial financing to modernize and expand began to develop. Banks insisted that loans be underwritten by continued high earnings. As traditional markets declined, manufacturers sought greater diversification in their product lines, phasing out less profitable ventures and finding new targets for investment. One such tactic was the creation of a broad, new market of currency speculation. In 1973 several developed nations agreed to remove their currencies from strict governmental control and to let rates "float" on an international market. The buying and selling of money became an important part of many companies' investment portfolios, but there was an unpleasant side effect. Currency speculation forced the U.S. dollar to record highs with respect to other currencies, thereby making American goods artificially more expensive and foreign imports considerably cheaper. Within this context of national and international developments, the domestic steel industry and Bethlehem in particular tried to stay alive.

Business History

In the early 1970s Bethlehem was still growing larger but not more profitable. Sales growth in absolute dollars was excellent, but profit rates were static. Furthermore, the company's markets were changing rapidly as key consumers such as the auto industry reduced their demand for steel. In the latter half of the decade, Bethlehem's orders declined 7.6% in their biggest market, transportation, and rose by only 3.2% in construction. Other markets remained stable. By classes of goods, the decline most adversely affected the company's largest lines — sheets, strip and tin (down from 34% to 32.8%) between 1977 and 1979, while it increased only slightly (12.1% to 12.7%) in bars, rods, tool steel, and semi-finished products during that same two years. These declines occurred just as the company was emerging from its first year of losses since the Depression (1977) and exacerbated the already-uncertain corporate policy decisions so battered by outside events.

Workers leave Lackawanna in the aftermath of a major shutdown, August, 1977. Protest sign in background expresses local sentiments.

By the early 1970s Bethlehem had eight steelmaking complexes in the U.S. The company had long pursued plans to expand by adding new facilities rather than upgrading existing operations. On one hand, when markets declined, they were burdened with overcapacity, but on the other they were mired by technological inefficiency. Furthermore, the slowdown in both backward and forward integration began to take its toll. Bethlehem was now forced to increase its expenses by purchasing supplies and services it had once owned. By 1979, the declining sources of raw ore had led the company to write 5-year contracts to buy 11.9 million tons of ore from outside suppliers. Bethlehem also had reduced the numbers of lake carriers it owned to transport raw goods and, with customers moving farther from the Great Lakes, became increasingly dependent on trucking to move their finished products, a transport service for which they had to pay outside providers. The instability of relying on such suppliers became evident in 1979 when 750 workers were idled at Burns Harbor and Sparrows Point due to a shortage of steel-hauling trucks.

Corporate Practices

The company tried to meet the changing world and transformed market demands with logical attempts to reorganize. They increased their bases of diversification and began to extend their investments overseas, both successful tactics of other corporations. At no point did Bethlehem ever reach the degree of diversification achieved by U.S. Steel in its take-over of Marathon Oil nor did Bethlehem's direct overseas investments ever amount to much beyond mining. But these tactics did draw off capital in non-steel directions, unfortunately without returning the hoped-for rewards, while enmeshing the company in numerous activities they were ill-prepared to handle to real advantage.

One of the first diversified acquisitions was the 1970 purchase, through a stock swap, of Kusan, Inc. A Tennessee-based plastics manufacturer with 10 plants in 6 states, Kusan did well enough in its lines of automotive, electrical, and other plastic products to acquire a subsidiary of its own from Bendix Corporation. A second acquisition by Bethlehem, also in 1970, was Multicon Properties, a housing developer. In 1973 the oil crisis prompted the steel company to enter oil and gas exploration, first to assure their own supplies and then for outside sale. In 1975 they entered a joint venture with Nuclear Dynamics called "Nubeth" to engage in uranium exploration and mining. Bethlehem held the majority 75% ownership in this operation.

Declining returns on Bethlehem's domestic reserves of ore and other raw materials led the company to try to upgrade its supplies through foreign operations. By 1972 the company's entire 72-million-ton supply of high-grade iron ore lay outside the U.S., with 31 million tons in Canada, 25 million tons in Liberia, and 16 million tons in Brazil. More than 20%, or 727 million tons, of their low-grade ores and taconite was also in Canada. Only the coal reserves lay exclusively inside the U.S. There Bethlehem did well, selling nearly 20% of their annual capacity on the open market.

Despite the attendant uncertainties of producing in overseas locations (both Chile and Venezuela nationalized Bethlehem mines in the early 1970s), the company expanded its foreign operations. In 1974 they opened another iron ore operation in Gabon, Africa and, through a subsidiary, invested $7 million in the activities of the international minerals conglomerate, Rio Tinto Zinc. Their boldest overseas operation began in 1978 when they started an iron ore mining and enrichment facility in Hopei, China. This was an extremely expensive endeavor and ultimately backfired since the new company sold iron ore to many Bethlehem competitors including Japan.

In a move combining overseas investment with diversification, the steel company began a new operation in Sierra Leone, Africa. Under direct Bethlehem management, the African operation would mine, process, and ship rutile,a titanium oxide used in paints, paper, and plastics. However, in 1982 Bethlehem was forced to sell its 85% interest to two other firms including Glickenhaus & Co., a West German corporation.

Raw materials were important, but by no means the only bases for overseas direct investment. In 1970 Bethlehem and a Singapore bank formed a joint 70/30 partnership called Bethlehem Singapore Private Ltd. They were to build offshore oil-drilling rigs to be sold in Hong Kong and China. However, in 1981 they stopped the manufacture and instead elected to sell the manufacturing technology rights to Wah-Chang International Marine Industries of Hong Kong and to Guangdong Shipbuilding in China.

The sale of technology is a sound strategy for generating income where the costs of production are too high, but it is a short-term solution at best. One of Bethlehem's best-selling product lines was its patented aluminum-coated sheet metal, Galvalume. Demand was good, but rather than shore up its shagging export operations by manufacturing Galvalume in the U.S. and selling it abroad, Bethlehem again sold the patent rights. In 1979 they licensed a Swedish firm, Svenski Stal AB, to produce and market Galvalume, then in 1980 licensed two more firms, Arbed and Phenix Werks, to do the same. The American steel company received the royalties for these operations — but lost all of the profits from either export or direct production.

At home Bethlehem was unquestionably dedicating large sums of capital to plant expansion, but the decision to increase capacity by creating entirely new facilities may have been counterproductive. A 1978 Brookings Institute study showed that new construction was far less cost effective than was upgrading existing facilities. New mills would undoubtedly reduce crew size and improve customer order turnaround time, thereby saving $60 per ton. But the capital costs would add $130 per ton over the costs of operating existing plants. Older mills which had already amortized their construction and incremental capital expenses were the only ones capable of operating on a cost-per-ton basis comparable with the steel mills of East Asia. So, Brookings concluded, new mills would not result in the level of savings anticipated.

Bethlehem had a strong resistance to revising their basic technology. With the company doing so well in the long post-war boom, there was an implicit belief that existing methods were sufficient. Experiments with continuous casting were abandoned as unworkable for large orders, leaving the Japanese to adopt and expand the process. When Bethlehem needed to catch up in the late 1970s, they were forced to write three costly technology-assistance pacts with countries which had surpassed them in continuous casting — Japan, West Germany, and Austria.

Bethlehem's strategies were not working. Their corporate planning began to reflect general business practice characteristic of the post-war slowdown. Rather than taking the long-range perspectives of a Schwab or a Grace, current management started making short-term plans in reaction to transitory phenomena. Although the 1971 employees' report declared that "Bethlehem will continue to take a hard look at every plant, mill, shop, product line . . . to make sure that each is pulling its share of the 'profit oar'," no one really knew what that entailed. Even though annual profits held remarkably steady, quarterly profits vacillated wildly. Company policy and practice turned on those three-month earnings, and decisions were made, then unmade, abruptly. In August, 1976 Bethlehem Chairman Lewis Foy declared Bethlehem was "fully competitive," but by September blast furnaces at the Sparrows Point, Maryland plant were closed. In February, 1977 6600 workers were laid off at Lackawanna, while in March the company aggressively recruited new management trainees for supervisory positions. In October, 1977 1800 were laid off at Cambria coal mining operations in Pennsylvania, then rehired 15 days later. In June, 1978 the company declared it had "healthy orders" but ended the production of forged railroad wheels at Johnstown that same day. Twenty-one days later they received an order for 1700 railroad cars from Trailer Train. They finally built a caster but did so at Johnstown where it did not fit. In 1977, when it was 90% complete, the $100 million project was abandoned. As we will see in Chapter 9, Lackawanna's mills opened and closed regularly. Managers and workers never knew, one day to the next, what their fate would be. Rapid stop-go planning in the quest for correcting short-term profits led one 30-year employee to observe, "They no longer made steel. They just made money."

Ownership and Control

The 1967 reorganization of Bethlehem's Board of Directors and top management had much to do with these new practices. By 1970 the company's top leadership resembled that of the old Lackawanna Steel Company. The firm once again was run not by men whose interests were in making steel but by those concerned with financial performance alone.

The new Board continued to attract leaders from top corporations as it had under Schwab. The Morgan connection, reconstituted after a 37-year hiatus, put directors from Morgan Guaranty in control of Bethlehem's policies as well as its debts. Chemical Bank, Bethlehem's largest creditor, also had a seat, while steel company President Stewart S. Cort continued his affiliation with Continental Illinois Trust whose stock represented a large part of steel company investments. On the other hand, neither Chase Manhattan nor Citibank, each with outstanding loans to Bethlehem, were represented on the Board.

In 1980, while no direct tie to Chemical existed, the relation with that bank was far from severed. Chemical's largest stock owners had long been the DuPont family. In 1980 Charles McCoy from E.I. DuPont de Nemours joined the steel corporation as a director. Morgan's ties were still active, though; former Morgan Chairman Ellmore Patterson sat on Bethlehem's Board and steel company Chairman Lewis Foy sat on the Morgan Board. Each had an interest in the other's company. Morgan had an outstanding 1977 loan of $300 million to the steel producer and would extend another $300 million in the early 1980s. Through the Morgan Board, Foy was able to meet regularly with his most valued customers, General Motors and Ford, as well as the old Bethlehem-Lackawanna ally, the Atchison, Topeka & Santa Fe Railroad. DuPont directors Howard Johnson and Edward Kane also served on the Morgan Board, making the Morgan-Chemical-DuPont-Bethlehem clique complete. While the numbers of "notables" now affiliated with Bethlehem were not as extensive as in the old days, they were nonetheless impressive.

The more important change at Bethlehem was not its renewed ties with its bankers but the internal reorientations of the company's own inside directors who comprised top management. An unbroken succession of steelmen had led from Schwab to Edmund Martin. His successors, however, represented a major departure from this long tradition. New chairmen Stewart S. Cort (1970-1974) and Lewis Foy (1974-1980) came from the company's ranks but through the marketing and purchasing divisions.

They were succeeded in 1980 by Donald Trautlein whose first exposure to the steel industry was auditing Bethlehem's books when he was a partner at Price, Waterhouse, the accounting firm. Before his election as Chairman, he had spent only three years at Bethlehem as comptroller. With a focus on the corporate balance sheet, he had little or no knowledge of production. Nevertheless, he was the architect of the 1977 shutdowns prompted by two bad quarterly returns — and accomplished on the eve of major new national rail demand.

The end of decades of leadership by steelmen sent perceptible shockwaves through the company. It was this new team which prompted overseas expansion and diversification on one hand, domestic shutdowns and layoffs on the other. The company now was sinking surplus cash into ever-greater portfolio investments so that its money and the workers' pension funds were invested in the stocks of its largest creditors and allies — Citicorp, Chemical Bank, J.P. Morgan (the private banking arm of Morgan Guaranty), and Continental. All of these banks, however, had lent $593 million to Japanese steel companies while squeezing the American steel producers fighting for scarce and costly capital.

Pressing for high returns, the company continued to construct new facilities — ignoring the Brookings Institute report — while shutting down existing operations even if they were still profitable. Rather than taking the long view which got Schwab through the Depression, contemporary Directors elected to put marginal facilities out of business. They took advantage of generous governmental cash rebates for rapid depreciation of existing facilities and added those funds to company profits. However, they underestimated the amount that shutdown would cost, so that roughly $750 million of the $800 million refunded for the 1983 closings at Lackawanna was absorbed by the costs of layoff and early retirement.

Financially-oriented managers clearly did not have the same perspective on company survival as did steelmen. Just as happy to make plastics as I-beams, they eliminated much of the company's productive base and therefore its hopes for recovery in better times.

Profitability

Despite company complaints about declining profits, Bethlehem made money every year 1970-1982, save for 1977. Gross sales remained high and were occasionally extraordinary. Between 1969 and 1973 sales jumped 100% from $2 billion to $4 billion. From 1973-1974, sales rose 20% to $5.4 billion, and by 1978 the company sold $6.2 billion with another $900 million increase in 1979.

Overall rates of profit remained steady for the period but did not grow substantially. As we have seen, the company's costs for supplies were relatively fixed; as sales increased, so did the expenses. The company profits got ever-larger in dollar returns but not in percentages of sales income. However, while annual profits were relatively stable and virtually always in the black, quarterly profits were quite volatile. For example, 3rd quarter 1973 earnings shot up 175% over the same quarter in 1972. In 1975 2nd quarter profits fell 29% compared to 1974, followed by a 3rd-quarter drop of 66%. But in the same two quarters of 1976 the company's earnings rose once again. After the disastrous losses of 1977, 1978 quarterly returns grew by as much as 288% in the 4th quarter. It was this yo-yo effect which kept company executives and their policies in a constant state of flux.

Rates of return 1970-1983 were definitely lower than much of the earlier postwar period, reflecting the general economic slowdown. Rates varied between 4.28% and 6.33% per year 1971-1975, declining in 1976 through the losses of 1977. However, discounting that one bad year, overall rates between 1976 and 1982 tended to remain at just under 4%. It was not until the last quarter of 1982 that Bethlehem began its steady decline and had actual losses. For the majority of the last years of Lackawanna's life, Bethlehem did not lose money — it just did not make as much as it was accustomed to. When they did lose money — and it was spectacular, $1.47 billion in one year — the losses came after they had closed Lackawanna and several other productive operations.

Prices

Unable to increase overall market demand, Bethlehem, along with most other American corporations, generated income by raising prices. Companies found that their steady customers would accede to these increases, especially if such costs could ultimately be passed along to consumers.

Between 1971 and 1983 Bethlehem raised prices 37 times on individual products or across the board. In 10 instances prices were temporarily rolled back or hikes deferred, but these were not permanent or long-lived reductions. In only one instance did the company respond to customer pressure when they rolled back prices on 15 commodities to meet demands from GM. Normally, however, prices rose between 6-15%, with some prices rising more than once in a calendar year. Even in periods of low demand and actual losses, Bethlehem raised their prices. In 1977 prices rose twice, once in February and again in June, for a total of 11%. A third 5.5% price hike for early 1978 was declared in December, 1977.

The decades-long industry agreements on pricing were never more evident. Virtually every broad increase was uniform among the major producers. U.S. Steel generally led the hikes, but in 1983 Bethlehem took control. The consensus on prices was at times so blatant that Bethlehem and several other firms were charged with price fixing by the federal government in 1974, 1977, and 1980 (a 1976 charge resulted in acquittal) and were charged with bribery to attract customers (rather than lowering prices) in 1976 and 1980.

Debt

Pressures on profits were the incentive to raise prices and came from two sources: debts and costs. In the 24 years from 1945 to 1969 Bethlehem incurred debts with a face value of $1.75 billion. In the 13 years between 1970 and 1983, the corporation added virtually the same amount, $1.72 billion. The company still had outstanding obligations from the first loan package worth, conservatively, $140 million (the figures are not public).

Bethlehem's largest debts were with their old, reliable creditors, Chemical and Morgan. Chase Manhattan still held a $70.5 million claim on the unpaid portion of their 1965 loan. As Bethlehem's business began to decline seriously, so did their credit rating, thereby making it more difficult to raise capital. The company was forced to turn to foreign financing to secure credit. In 1983 they raised $540 million from three separate banking groups, including one comprised of Austrian banks, in order to finance the technology-assistance pact they had drawn with Voest-Alpine, an Austrian corporation, to install two continuous casters.

The debt package put unprecedented strain on corporate resources. Throughout most of the 1970s and early 1980s interest rates soared to record highs; the double-digit interest not only increased the direct claims on corporate earnings through bond and bank payments but also hurt consumer sales in such areas as autos, thereby lessening demand for steel.

The company's debt/asset ratio climbed precipitously. While consumer loans were pegged to monthly earnings, corporate debts were much larger and were set in relation to overall holdings. In 1973 Bethlehem's ratio of debts to assets was 23% but by 1977 climbed to 35%. The debt-to-asset ratio diminished somewhat in 1979, but soared to 46% after the shutdown of Lackawanna in 1983. Thus the closing of "unproductive" facilities had the unpleasant effect of saving expenses on one hand but removing assets available for securing loans on the other. In 1960 the company had $3 for every $1 owed, but by 1980 they owed $4 for every $1 they had on hand.

Costs

The same costs which were part of Bethlehem's expense package in the 1960s were still present in 1970-1983; the costs of supplies and of capital have been noted above. The cost of labor, however, had several new factors with which the company had to contend and which excited considerable concern for the company and its executives.

Wage costs were still below 40% of the sales dollar in the early 1970s. As part of the national negotiations between industry and the United Steelworkers of America, the company had exactly the same increases in pay as all other steel majors. In 1974 the USW proposed and passed a "no-strike" clause in exchange for a 3% annual increase, a one-time bonus of $150, and cost-of-living increases tied to the rate of inflation. The dollar value was, at the time, not particularly onerous to the industry; no one expected the return of double-digit inflation, and the wage increase was well below the rates of increase in prices.

The union pegged demands to its perceived understanding of the industry's profitability, widely touted in the press by Bethlehem's executives as well as those from other corporations. Companies refused to "open the books" to employees, fearing an invasion of corporate power, which might lead "insidiously" to "joint decision-making". Rarely did the company's public reports indicate dissatisfaction with rates of return. The 1977 and 1980 labor contracts retained the basic provisos set in 1974, but by this time the inflationary spiral was beginning again. COLA costs were becoming a critical expense, but the very inflation which undermined the purchasing value of steelworkers' income also inflated Bethlehem's prices and improved its income. Neither side was willing to reduce demands; it was a vicious circle from which no one was emerging a winner.

There were factors which ameliorated some of Bethlehem's labor costs. A combination of plant improvement and human effort increased productivity. Between

1960 and 1969 the company's annual production of ingots and castings averaged 145.8 tons per employee. In the 1970s, however, not only did the average production rate rise to 175.9 tons per worker but it met or surpassed 200 tons/person in 1973, 1978, and 1979. These were the highest per capita figures in the company's history.

Other factors, however, conspired against Bethlehem and the steel industry and served to inflate labor costs as economic and political factors in the global market. American wages historically had always been high. That was one factor contributing to the nation's strength as a high-consumption as well as high-production country. By the late 1970s and early 1980s, however, as demand slowed and profits were squeezed, the high value of the dollar on the world market seriously undermined U.S. industries. Relative to all other nations, American labor became "overpriced" due to the discrepancies in currency valuation, and the whole nation began to demand worker concessions. American goods became relatively more expensive, and the domestic market soon was flooded with imports replacing American manufacturers. Had there been no pay increases at all, this still would have occurred. Steelworkers were understandably reluctant to give back wages or benefits which bought no more on the domestic market. In 1983 when the industry was clearly in trouble, the union did concede 9% reductions. This made little difference internationally, however; the dollar climbed still higher, thereby offsetting the union giveback and continuing to make U.S. steel much more expensive than that of other nations.

Management costs were also a problem for Bethlehem. In the boom years following the war, the numbers of management personnel had increased substantially. Middle-level and top-level positions were increased as operating divisions were multiplied and support staff added. A 1977 move to reduce staff by laying off 2500 did not last; virtually every position was refilled within a year. Also in 1977, six top officers received raises in their base salaries. When Trautlein did make drastic cuts in the early 1980s, it became clear that the sheer cost of these layoffs virtually exceeded their gains. It could cost as much as $2 million to phase out a single vice president.

Layoff costs were a new type of labor expense which increased corporate liabilities with absolutely no return in productivity. There was an ever-growing pool of men and women with claims on corporate resources for short- or long-term layoffs and retirement. In 1977 the layoff of salaried workers under early retirement cost the company $36 million. Extended unemployment benefits were allowed to the blue-collar workers let go in 1977 since their dismissal was due to foreign competition. The 1983 shutdown of Lackawanna cost Bethlehem $55,000 per worker and eroded the tax rebates from the federal government, eating up almost the entire $800 million. The company tried in 1982 to slash benefits in health insurance for its retirees, but in 1984 a federal court mandated restoration of those payments so that the company continued to carry this burden. Clearly, going out of business was an expensive proposition.

At exactly the time Bethlehem's rates of return were declining, the company's strategy for generating stockholder support was predicted on increased dividends. The steel industry as a whole paid our 43% of profits to shareholders, well above the national corporate average. Between 1975 and 1980 Bethlehem paid out 83%. The company's per-share payments never fell to less than $1.00, even when earnings were low, and not even in 1977 when there were losses. With 80 million shares of stock, such expenditures clearly eroded the company's cash base. Rather than pursuing Schwab's strategy of withholding dividends in lean times to strengthen the company, Bethlehem's current management elected to promote the fiction of corporate well-being by distributing precious reserves, thereby leaving no cushion to support the company through hard times.

In June, 1983, the *Buffalo Evening News* wrote that the "strength of Buffalo's steel industry is melting away." Buffalo was once a "rival of places like Pittsburgh and Youngstown and Chicago as an iron and steel center" but now "the sun sets on Buffalo's iron age." Lackawanna's steel plant ended much as it began, the jettisoned facility of a powerful corporation. Once again cost-benefit analysis had

declared the operation redundant. Still capable of producing steel, it was no longer capable of generating sufficient profit. Victim not once but twice of financial calculations coupled with benign neglect, this time Lackawanna went out of the steelmaking business forever. Sixty-one years to the month after the once-coveted facility had become part of Schwab's mammoth steel empire, it was all over. The fires were banked, the mills stopped rolling, and thousands of men and women walked out of the gates, never to return.

8. THE PLANT

In 1970 Bethlehem Steel diagnosed the Lackawanna Plant as having cancer, but the patient took 13 years to die. The plant's decline was in direct contrast to the heightened activity of the late 1960s, but the sequence of partial shutdowns extended the economic, social, and psychological traumas by prolonging the climate of uncertainty rather than concentrating emotion at one cataclysmic closing. Lackawanna expired in pieces as Bethlehem lopped off facilities considered economically marginal.

Bethlehem first announced a cutback in rated steelmaking capacity from 6,700,000 to 4,800,000 tons on 30 November 1970; plant capacity was slashed again on 18 August 1977 to a nominal level of 2,800,000 tons. On 27 December 1982 Bethlehem announced that it planned to terminate all steelmaking during 1983. With each reduction in steelmaking there were related cuts in coke oven, blast furnace, rolling mill, materials handling, maintenance, and transport operations. By the time the BOFs grew cold after October, 1983 many other parts of the plant had already received last rites.

BLAST FURNACES
Traditional conditions in cast house (1983).

Lackawanna's decline coincided with Burns Harbor's maturation. When Bethlehem completed this newest plant in 1970, some contraction at Lackawanna became inevitable. As one steelworkers' union official noted:

> The Burns Harbor plant still has bugs in it, but once they're ironed out — as they inevitably will be — the production possible at that mill will be tremendous.
>
> The result will be a highly competitive situation within the Bethlehem system itself — among its various mills — with the automated, sophisticated Burns Harbor plant siphoning more and more production away from other, older mills in the company system.

Within the domestic steel industry new patterns of raw materials flow and product distribution downgraded the strategic position Lackawanna had occupied when it began as a runaway plant from another state in 1899. Evaluating a mature works that had grown by accretion over several decades, Bethlehem directors faced the same choices as turn-of-the-century Scranton steelmakers about whether to incorporate new capital and technology into a congested plant or abandon it outright.

By the late 1960s some of the factors that had once made Lackawanna a prime location for raw materials assembly and product distribution were no longer as powerful. Beneficiation techniques such as pelletization still sustained the shipment of lower-grade iron ore from the upper lakes. However, during the post-war era companies had devised an even more extensive global mining and transportation system that could deliver high-grade ore from Liberia or other foreign nations to tidewater ports year round. Lakeside plants derived less direct benefit from the new worldwide ore supply network because of winter interruptions to navigation on the St. Lawrence Seaway as well as congestion at the Welland Ship Canal. Alternative sources of domestic high-grade ore and new transportation methods also came into play. For example, Bethlehem planned to bring ore from developments near Pea Ridge, Missouri, to Burns Harbor by rail. Metallurgical coke supplies remained available in the traditional areas of the northern Appalachian plateau, and a spurt

STEELMAKING
Modern BOF computer control room (1981).

of innovations for improving blast furnace fuel economy reinforced the continuing decline of steelmaking districts whose primary advantage had been proximity to the coal fields. Fuel sources did not affect Lackawanna's position materially, but the new ore supplies put the plant at a slight competitive disadvantage with some other sites despite its lakeside location.

Lackawanna's marketing situation had deteriorated more drastically. Exports to Canada declined. Deindustrialization in the Northeast and the continuing westward shift of steel markets meant that many of Lackawanna's potential customers were now closer to other plants including Burns Harbor; in fact, Lackawanna's share of the automotive market had declined 20% from the late 1950s to the late 1960s. Following World War II the major automobile companies had shifted many of their component and assembly operations out of the Detroit area. While Ford's new stamping plant near Lackawanna was one outgrowth of decentralization, a more substantial portion of these facilities still remained within the western portion of the Manufacturing Belt that now lay inside Burns Harbor's orbit. The midcontinental marketing zone, once a major tributary of Lackawanna, shrank to include only the areas east of Detroit. Within this truncated midwestern market Lackawanna continued to try to dispose of 94% of its output; only 6% went to New York State.

Changing modes of transportation also operated to Lackawanna's disadvantage. Water shipment had been suitable for structurals which could be ordered well in advance but was not satisfactory for flat-rolled sheet and coil products since considerably more delivery time and product rehandling were involved. As late as the 1950s Lackawanna had delivered over 1,000,000 tons to Detroit in lake ships, much as Eugene Grace had forseen in the 1920s; in 1970 Bethlehem sold its last crane boat used for lake shipments of finished steel. Many large customers came to prefer truck deliveries direct to their own warehouses. However, Lackawanna gained no inherent advantage from overland shipments to compensate for the loss of water deliveries. The plant was further from Detroit or Chicago than a number of other mills, and the strip mills produced coils of an awkward size that complicated the economics of making up truck loads. As in the case of raw materials assembly, Lackawanna's inherent marketing advantages of sixty years earlier had become moderate disadvantages in an increasingly competitive environment.

Record operating levels during the late 1960s exposed some of the areas where the Lackawanna plant no longer measured up to the best modern practices. Producing steel for Burns Harbor as well as its own internal consumption, the plant responded like a middle-aged athlete suddenly thrust into Olympic competition; it became apparent that some of its strengths had deteriorated. The new BOFs gave Lackawanna a bionic heart, but older departments had to match its pace without equivalent modernization.

The utility systems that circulated steam, air, water, gas, and electricity through the plant were showing signs of arteriosclerosis. Shortages of steam for turbo-blowers and turbo-electric generators occasionally hampered operations. A number of air compressors verged on the antique, and spare parts had to be fabricated at the plant since they could no longer be purchased. Reserve pumping capacity was slim, and water from the ship canal was not an ideal source for cooling or condensing purposes. Electrical generation and distribution were complicated by the coexistence of two different frequencies, since a program begun in 1949 to replace 25-cycle facilities with 60-cycle equipment was not yet completed. Utility equipment was dispersed in relatively small units throughout the plant rather than being concentrated in larger units at central locations. A proliferation of pipelines and other redundant distributing systems represented additional residues from the plant's stages of growth.

Lackawanna Steel's original layout of parallel material and process flows had integrated the sequences of plant operations with exceptional rationality. Over the years, piecemeal expansion south of the original plant or across the Hamburg Turnpike meant that new materials handling routes, production facilities, and product handling needs had to be synchronized with older facilities.

By the late 1960s some recent technological advances had been grafted onto certain portions of the plant, but Charles Schwab would still have recognized other

STEELMAKING
BOF interior. (L) vessel charging platform. (C) main and auxiliary hoists of 350-ton charging crane (1983).

equipment. It was a time of transition in the steel industry. An older generation of capital investments had reached a plateau; corporations earned diminishing returns from substituting capital in the form of conventional processes for labor; higher capital costs threatened to cancel out any savings from reduced operating costs. New techniques for making and shaping steel such as oxygen blowing, continuous casting, and process computerization held out the prospects of raising productivity and profits. However, this new pool of innovations was already available in foreign-owned as well as domestic plants, giving progressive firms overseas a decisive productive edge.

Some engineers dreamed of new concepts that could convert steelmaking into more of a continuous flow process with fewer cycles of expensive heating and cooling. But, few of these visions would be implemented at Lackawanna. The plant's increasingly marginal position within Bethlehem's overall corporate plan meant that, for economic and political reasons, other works had higher priority for capital allocations. While technological change did not come to an immediate halt at Lackawanna after 1970, significant expenditures were confined to a few projects that proved insufficient to stem the continuing erosion of overall capacity.

Coke Ovens

Since improved fuel economy in blast furnaces continued to diminish coke consumption rates, there were no radical innovations in oven design. Layouts of modern coke plants favored taller ovens grouped into as few batteries as possible to reduce labor costs while facilitating coal and coke handling. Bethlehem had scheduled a program that would have built three new batteries to replace eight existing batteries over the period 1969-1973. However, only one of the new batteries (#9) was ever constructed. The eight older batteries had rated capacities of 700-1170 tons per day; the daily capacity of the three replacement batteries was to have been 2427 tons of coke each. The 6-meter ovens in #9 battery were comparable in size to installations at Burns Harbor.

STEELMAKING
BOF charging platform. (L) vessel tilted down. (C) scrap charging apparatus (1983).

Recovery of gas and coal chemicals from by-product coke ovens had been a key element in the original design of the Lackawanna steel plant. However, in the late 1960s the economics of by-product recovery were different, and utilization of by-product plant capacity had changed as a result. Coal chemical processing at Lackawanna included facilities for recovery of light oils and ammonia as well as removal of sulphur from coke oven gas. At modern plants such as Burns Harbor more light oil was left in the gas to enrich its btu content for use as utility fuel. Furthermore, Bethlehem considered that refining light oil in competition with large oil companies was a losing proposition, so Lackawanna had ceased operating that portion of its facilities during the early 1960s. Recovery of ammonium sulphate from the gas and flushing liquor was likewise judged to be uneconomical. Removing sulphur from coke oven gas was critical when it was used as fuel in the open hearths, but replacement of open hearths with BOFs would eventually render the process redundant. By comparison with the new facilities at Burns Harbor, Lackawanna's by-product plant was thus considered more costly to operate, due in part to its excess capacity for recovering chemicals which were no longer desirable.

Running the new #9 battery in harness with older portions of the by-product plant had the same effects of exposing defects in outmoded equipment as in the case of the BOF. With the 6-meter ovens in operation, overall gas production was up, but deriving full benefit from greater output would entail investments in equipment to bring gas handling, flushing liquor, and cooling water capacity up to the same level.

Coke plant operations fluctuated with the overall pace of plant activity during the early 1970s and then shrank in proportion to reductions in pig iron and steelmaking capacity. The #3-#6 batteries, dating from the 1920s, were relegated initially to reserve status but saw some service during periods of peak demand. The 1977 cutback projected that 5 out of 7 existing batteries would be eliminated, leaving only #8 and #9 in operation; however, #7 battery came back on line after extensive repairs during 1979.

ROLLING MILLS
Norton-Centro-Maskin grinding machine replaced swing-frame grinders for conditioning billets by removing surface defects (1973).

Emissions from the coke ovens were a focal point of friction with environmental agencies. Bethlehem grudgingly allocated some funds for investments which it characterized as controlling pollution rather than increasing production. In 1979 a one-spot coke-pushing emission control car was installed at #7 and #8 batteries to cleanse the gases emanating from hot coke after it had left the oven. A similar mobile unit was planned for #9 battery as well, but delays resulted when the equipment supplier encountered business problems. By 1980 both #8 and #9 batteries required extensive rehabilitation. Number 9 battery should have lasted 25-30 years, but Bethlehem shelved plans for a complete rebuild of this key battery in 1981. The remaining coke ovens were spared during the 1983 shutdown. Number 7, #8, and #9 batteries continued in production, but their coke went to other steel plants.

Blast Furnaces

Improvements in blast furnace construction and practice as of the late 1960s included new means of accomplishing some of the same goals that pig iron producers had pursued in the late 19th century. During the 1880s, Capt. Bill Jones and Julian Kennedy had boosted the driving rate of the furnaces, and stingier South Chicago practices had improved fuel economy. Some 80 years later current methods for attaining faster driving and reduced coke consumption included operating furnaces at higher top pressures (a concept first implemented during World War II); beneficiation of ores and coke to improve iron content or particle size; supplementing the blast with hydrocarbon fuel injection or oxygen enrichment, and regulating the moisture content of the blast to achieve higher temperatures. As a result of improved practices, the optimal size of furnaces was growing and the number of units required was decreasing. Some computerized control systems were introduced to monitor top gas pressure and blast moisture variations or to coordinate the charging of materials with other furnace operations.

The furnaces and support facilities at Lackawanna fell short of these standards in some respects. The seven existing furnaces had daily capacities ranging from 1200 to 3200 tons; modern furnaces at other contemporary U.S. plants averaged in excess of 4000 tons per day, and some Japanese furnaces were designed for up to 6000-7000 tons per day. Lackawanna's stacks were also smaller than current designs with hearth diameters ranging from 21' 3" to 29' 11" compared to an industry standard of 35'.

Some features of the blowing system that supplied air for the blast were outmoded. The six centrifugal turbo-blowers were not as flexible in operation as more efficient axial units with adjustable vanes and diaphragms, and they suffered from occasional steam shortages as well. Eleven gas engines provided reserve capacity but required maintenance of an alternative fuel distribution system that paralleled the steam supply for the turbos; the benefits of recycling waste gas were now considered less economical than at the turn of the century. Total usable blowing capacity amounted to 823,000 cfm at 25-35 psi with 625,000 cfm from the turbo-blowers. Greater overall blowing capacity, concentrated in a smaller number of larger units, could have raised pig iron output and reduced operating costs. All the furnaces featured natural gas injection and had provisions for oxygen enrichment to increase production without additional air blast volume. However, the availability of oxygen was limited. Blast temperatures were limited to 1100-1900^{0} compared with a best-practice level of 2000-2200^{0}, leading to coke consumption rates that were higher than optimal.

At the support facilities, the dockside ore storage yard had reached capacity, and existing stockhouse equipment limited the rate at which furnaces could be supplied with materials. The faster heat times of the BOFs and their greater appetite for hot metal exacerbated existing bottlenecks in materials handling and pig iron production.

When Bethlehem first projected what Lackawanna's steelmaking requirements would be after it ceased to supply Burns Harbor, it seemed likely that two or three blast furnaces would be shut down permanently. However, during the early 1970s the occasional revival of open hearth steelmaking in periods of peak demand raised the plant's overall requirements for hot metal. As a result, idle blast furnace and coke oven capacity was dusted off and put back in temporary service. Even B furnace emerged from retirement in 1973, despite being handicapped by museum-vintage equipment and the lowest capacity in the plant. The closings announced in 1977 included mothballing 4 of the remaining 6 blast furnaces, leaving sufficient hot metal capacity to supply two BOFs. H and J furnaces were the ones slated for survival, but C furnace continued operating into 1979 as well. The sintering plant also remained active; it was modernized in 1979 and equipped with new pollution controls as late as 1981-1982. Lackawanna's pig iron production ended with the last cast at J furnace in October, 1983.

Steelmaking

Bethlehem was to have implemented the 1970-1971 cut in Lackawanna's rated steelmaking capacity by phasing out all the surviving open hearths and concentrating production in the BOF shop. By comparison with the BOFs, open hearths took longer to make a given heat, required more fuel, and were controlled by the individual judgment of the first helper rather than computers. Age, physical deterioration, and functional obsolescence also counted as strikes against the open hearths.

Diversion of hot metal to the BOFs hampered open hearth operations. The proportion of molten iron used in the average open hearth charge dropped from 60% to 25-35%. As a result the furnaces had to melt larger cold charges which resulted in lower yields, more damage to refractory linings, and more frequent down time for rebuilding. With the advent of the BOF, utilization of available open-hearth capacity dropped dramatically during the late 1960s from levels of over 90% to a range of 30-52%.

In 1968 #2 open hearth had a yearly capacity of 1,200,000 tons from 9 basic furnaces. Three of the furnaces were fitted with oxygen roof lances, and six were equipped with electrostatic precipitators for emission control. Although vacuum degassing gave the shop the capability of producing cleaner steels in high-quality heats, the original material and product handling layout was not well suited to specialty operations. Stocking, charging, and teeming aisles were narrow, and only a single track led from the stockyard to the charging floor. Heavier cold metal charges required more scrap buggies per heat and led to congestion in the charging aisle. Specialty heats with a number of smaller ingots could clog the teeming area, and space limitations affected the rhythm of pouring.

Number 3 open hearth shop was also not without problems. Continuing foundation settlement had resulted in distortion of structural supports and equipment connections. Number 3 open hearth contained 11 furnaces; all furnaces except one were lined with silica refractory bricks. Six had oxygen roof lances and emission control devices. Production capacity was rated at 1,500,000 tons, mainly of standard quality steel for structurals and slabs, the type of order that was being shifted over to the BOF. As at #2 shop, environmental regulations precluded operation of units that lacked pollution control equipment. Both #2 and #3 shops were idled during Lackawanna's initial downsizing in 1970-1971. However, several furnaces in #2 shop were revived during the 1973 boomlet that also marked the last campaigns of older coke ovens and blast furnaces.

Bethlehem did direct substantial investment into Lackawanna's BOF shop during the mid-1970s. An R-H vacuum degassing unit was added in 1973. Early in 1974 a program was announced to replace the original BOF vessels with larger 300-ton models, and experimental scrap preheating was tried as an attempt to reduce the amount of molten iron that had to be squeezed out of the blast furnaces. Another new venture along the same line was the scrap melter built during 1975-1976. This new facility was designed to furnish the BOFs with an additional 800,000 tons of molten metal per year from in-plant and purchased scrap, but it was initially plagued with electrical difficulties. In 1977 one of the three BOF vessels was scheduled to be sidelined when plant capacity was cut, but during the late 1970s operating levels remained higher than Bethlehem had anticipated and BOF shop capacity held at

a level of 3,500,000 tons. The lime plant that furnished flux was taken out of service in 1981. In October, 1983 the BOF shop poured its final heat.

Rolling Mills

When Bethlehem cut Lackawanna's steelmaking capacity in 1970, its announced purpose was to bring the plant's ingot production back in line with its potential finished product output. However, during the late 1960s Lackawanna's rolling mills were operating below their nominal capacity. Utilization rates ranged from 35% in the structural mills and 50% in the bar mills to 65% in the hot strip mill, and 70-75% in the 40"/30-21" and 44"/32" complexes. These figures indicate some erosion in the plant's marketing position as well as various degrees of obsolescence in the mills themselves.

For example, in the late 1960s, the 79" hot strip mill played a critical role in meeting flat-rolled product demand. However, it was basically an updated first-generation strip mill that had to compete with other mills of modern size and design, including Burns Harbor. Lackawanna's five slab reheating furnaces were rated at 92-100 tons per hour compared to the 115-ton per hour rating of modern units; modern facilities also featured improved zone heat control, and the furnaces at U.S. Steel's new mill in Gary were equipped with mechanical fingers that deposited the heated slab on the roller line to minimize damage. Other strip mills also had a considerable advantage over Lackawanna in horsepower per stand, maximum delivery speed, and maximum coil weight.

Advances in flat-rolled quality control during this period were even more impressive. Meeting more exacting customer specifications, particularly from the auto industry, required precise control of temperatures and drafts when rolling particular grades of steel. Working the product at the correct temperature for the desired amount of reduction resulted in more uniform grain structure, hardness, and ductility. The steel industry pursued these quality improvements by investing in new equipment that once again altered the work process in hot strip mills, much as continuous rolling had crowded out hand sheet mills in the 1930s. Integrated data logging systems and computer controls supplemented or superseded the judgment of operators in making corrections for variations in temperature, speed, or reduction.

Other processes coming into vogue during the late 1960s never made an appearance at Lackawanna but did serve as yardsticks to measure how far older methods were falling short of the best practices at other plants. For example, continuous casting of semi-finished slabs or billets direct from molten steel held out the prospect of eliminating both steps and jobs in the traditional method of ingot reduction, including teeming, stripping, soaking pits, and primary rolling mills.

Bethlehem's major investment in improved finishing capacity for Lackawanna revolved around construction of a new 13" bar mill. The project was announced in 1972 as the corporation's part of a complex tax reduction and labor productivity agreement with local governments and the United Steelworkers union. Intended as a replacement for the 1947 10" mill, the new computer-controlled facility would be housed in one of five new buildings tacked onto the existing bar mill complex. Parts of existing structures would be used for billet storage, finishing, and shipping, but new equipment would include a reheating furnace, cooling bed, annealing furnace, shears, bar coiler, and coil pickler. The adjacent 12" mill was expected to remain in operation. Bethlehem also undertook construction of a new reheating furnace and transfer equipment at the 44" and 30-21" mills that were to supply the new bar mill with billets. This revised billet rolling set-up with its new recuperative walking-beam reheating furnace was operable by 1975. The 13" mill initially was projected to be in production by the end of 1974, but it was not actually ready until early 1976. Designed by the German firm of Demag, the 2-high, 22-stand mill featured 9 computers.

As with the galvanizing plant during the 1960s, the new bar mill's impact on employment levels was expected to be felt as much from increased operating levels in steelmaking departments as from new jobs in the highly-automated facility.

ROLLING MILLS
79" mill runout table and water sprays for cooling continuous strip between finishing stands and coilers (1982).

However, like the BOF and #9 coke battery, the 13" mill was another jewel in a rusty crown. Juxtaposing advanced technology and older operations within an integrated steel plant underlined the shortcomings of outmoded sectors even while attempting to revive them. Nevertheless, the 13" mill did represent Bethlehem's final attempt to hold Lackawanna's market for automotive and forging bar stock against the non-integrated domestic mini-mills or imports from foreign corporations that had modernized their plants even more extensively.

The dramatic 1977 cutbacks brought down the curtain on several mills. As a prelude, Bethlehem announced in early June that the 28" mill would close at the end of July. The events of 18 August were obituaries for the 54-48", 40", and 32" mills along with the hot strip mill finishing department. Thus, the rail products that had given birth to the original Lackawanna plant and the structural products that had once helped build Chicago, all passed into history.

The last act was played out after 1980. The 12" finally expired in 1981, and in 1983 the 45" x 90", the 79" hot strip mill, and the 44"/30-21" all rolled their last steel. The 13" bar mill has stayed in operation, using billets made in Bethlehem's Johnstown, Pennsylvania plant, while the cold reduction mill and galvanizing plant continue working on coils shipped in from Burns Harbor. But in the remaining empty mills, only darkness and silence prevail.

9. WORK

ROLLING MILLS
Last turn in the 79" mill. The crew set an all-time record by rolling 275 slabs (1983).

As Lackawanna entered its last summer of steelmaking in 1983, a Buffalo reporter assigned to the death watch interviewed several retirees about their experiences in the plant. The conversations inevitably turned toward the reasons for Lackawanna's demise, and the older men identified several changes they had noticed over the years.

Joseph Kozak, who had begun working at the plant in 1935, pointed out shifts in the types of steel demanded by customers. He maintained that "We could sell plenty of steel if car manufacturers went back to heavy gauge." John Walczak, who had exchanged a spot in a Civilian Conservation Corps camp for a job at Lackawanna in 1940, recalled that over the years the plant had become fragmented into "little kingdoms" of management. Al Sunick, a former vice-president of USW Local #2601, criticized Bethlehem for keeping Pennsylvania facilities, as antiquated as Lackawanna, on life-support systems while pulling the plug on the New York plant. Several of the men praised Edmund Martin as the last general manager and corporate executive who identified personally with Lackawanna and believed the facility could remain strong; workers believed that his successors all perceived the plant as an expendable facility and their time spent at Lackawanna as just another step up the corporate ladder. Retirees Victor S. Giancaterino and Edward Rybak touched on the long shadow cast over Lackawanna by the creation of Burns Harbor.

Over the years these men had watched the plant's employment fluctuate and dwindle from 21,500 in 1965 and 18,500 in 1969 to 12,000 by 1972-1973, 8500 by 1977, and 7300 (including 3400 on layoff) in late 1982. Pessimistic in the face of continuing cutbacks which culminated in shutdown, 31-year veteran Frank Gombos summed up his feelings by remarking, "Now I wait for the afternoon paper to see what the next blow has been."

In their recollections the retired steelworkers gave little weight to the factors conventionally cited as contributing to the plant's demise. Since 1970 the local and national media had devoted reams of copy and segments of air time to cataloging Lackawanna's woes. Bethlehem took the public position that the stability of the plant was being eroded by disproportionately high local taxes, unrealistic environmental regulations, and declining labor productivity.

ROLLING MILLS
Last turn in the 79" mill (1983).

But, as the retirees' silence on these points seemed to indicate, these ills were symptoms of causes much larger than a single plant, town, company, or even industry. Thorny issues of taxation, pollution, and labor relations would not have assumed such critical proportions without the global transformations affecting U.S. corporate policy that were discussed in Chapters 4 and 7. Despite corporate claims that Lackawanna city taxes comprised nearly a third of the plant's steelmaking costs (but which were fully deductible from federal taxes), the tax and environmental issues were more significant as occasions for Bethlehem's exercise of power. The corporation exploited the constitutional legacy of federalism by pitting separate communities and states against one another in competition for scarce resources, tying the continued existence of their operations in any given community to civic compliance with corporate demands. Inside the plant demand for local concessions on work rules not only revived labor's legal and historic disadvantages with regard to management's authority but also exposed some of the contradictions in the post-war patterns of industrial relations.

Since the 1940s collective bargaining had regularized settlement of workplace conflicts without removing the bases of that conflict. Post-war labor legislation and bargaining practices codified the same rigid dichotomy between the roles of managers and workers that was practiced during the birth of the Lackawanna plant. As noted in Chapter 6, the authority to operate the business and direct the work force lay exclusively with management. As educator Jack Metzgar has written:

> Whether and where a company does business, the overall direction of an industry and the condition of the economy are some of those larger decisions from which the working class has largely excluded itself. Shutdowns and cutbacks are not negotiated. They are announced.
>
> . . . the union can negotiate the terms of disemployment but it cannot by the rules of the game, challenge the decisions to shut down — *or any of the decisions that have led to the decision to shut down.*

In return for abdicating involvement in corporate decision making, organized workers secured some protection from arbitrary managerial authority and obtained a rising standard of living. The cutbacks of the 1970s fell on this established machinery of labor relations from an earlier era; the post-war social compromise now took a new turn as management began reasserting its authority to alter the terms of the bargains struck between the 1940s and the 1960s.

The basic conflict over authority within the workplace became evident as early as 1970. In September of that year 30 men left the 44" mill in a spontaneous walkout reminiscent of the early 1940s. The issue was a company initiative to eliminate jobs by combining the soaking pit operations at both the 44" and 40" mills. The protest against management's attempt to implement this change without negotiation led to a work stoppage at other departments as well, and the company obtained a court injunction against the job action on the grounds that it violated the existing contract. The men did return to work, and two weeks later management and USW Local #2603 took the question of job elimination into arbitration that dragged on through July, 1971.

This was a scenario that might have been played out in the militant days of the CIO, but such episodes took on a different meaning in the economic and political context of incipient deindustrialization. The dispute in the 44" pits foreshadowed a series of conflicts over the "past practices" clause in the section of the union contract dealing with work rules under which operations were performed or manned. As Bethlehem's 1970 plan for shrinking Lackawanna became reality in the fall of 1971, leaders of Local #2603 and other USW bargaining units believed that excessive job combinations and eliminations were abuses of the union's recent concessions regarding increased productivity. Now however, contractual provisions had curbed wildcat strikes; the union's only recourse was the grievance procedure. Over a 2-week period in October, workers filed 300 grievances protesting the most recent job combinations.

Some steel industry observers have singled out union rigidity over past practices as the root cause for the decline of the U.S. steel industry in general and Lackawanna in particular. In December, 1970 Bethlehem Vice President John Jacobs, a former manager at Lackawanna, noted that overstaffing and the inability to reduce crew sizes hampered management efforts to improve what the company perceived as poor productivity at the New York plant. Using Burns Harbor as a yardstick, Bethlehem calculated that the newer plant could produce 4,300,000 tons of steel with only 6400 employees, but Lackawanna required 12,000 employees to turn out its truncated capacity of 4,800,000 tons. Part of the discrepancy was traceable to Burns Harbor's modern equipment upon which Bethlehem had lavished over $1 billion in capital expenditure, and the Indiana plant's concentration on flat-rolled products made meaningful comparison between the two facilities difficult. However, it was also noted that over decades of operation Lackawanna had accumulated more work traditions than Burns Harbor where "The work force has no past practice clause since there were no past practices before 1966."

While the past practice clause originally inserted in the 1956 USW contract may be construed as a barrier to meaningful changes mandated by globalized production pressures, it is also true that the language represented a standardization of procedures for insuring that any new job structure would be the product of mutual negotiation rather than arbitrary managerial decisions. Like so much of post-war collective bargaining, past practices represented labor's traditionally defensive orientation in the face of management's greater legal authority.

There were some efforts to seek compromise solutions to Lackawanna's problems. In June, 1971 Local #2603 staged an unprecedented mass meeting among company officials, supervisors, employees, and union representatives to resolve the festering situation in the 44" mill soaking pits. Later, in October, 1971, Art Sambuchi, Local #2603 Chief Grievance Officer, chaired a meeting at which a union fact-finding committee supported Bethlehem's contention that some relief on tax and environmental issues was necessary. Concessions on both issues were secured for the company, netting a 24% reduction in local taxes and significant delays on implementing pollution-control measures. In 1972 labor's cooperation over cost-saving productivity issues was instrumental in securing Bethlehem's commitment to construct a new bar mill at Lackawanna. But the local took a hard line on defending their legal right to negotiate over job restructuring, playing their traditional defensive role in the divisions of authority over the workplace. Bethlehem, compared to other corporations, was slow to create joint labor-management participation teams that might plan future production cooperatively. The first such team was developed in Johnstown in 1983 after Lackawanna's fate was already decided and shutdown well underway.

Grievances over managerial modifications of past practices also involved safety issues. Regular spells on hot jobs and sufficient crew size for production and maintenance operations to guarantee worker safety made a critical difference in health as well as longevity on the job. The relation between safety violations and job cutbacks was a key issue in the 1971 industry-wide contract talks about past practices.

Even with the safety equipment, working in a steel plant was still hazardous. Mechanization had eliminated much of the back-breaking labor over the years, but the grime, dangerous gases, dust, and brutal temperatures still remained. Coke oven operator Mike Reilley told journalist Rachel Scott in 1974 that "They give you a respirator. It stops the bigger chunks . . . but the smoke and gasses go right through." Frank Blakeley, Jr., another coke oven operator, echoed the same concern a few years later. Johnny Quider recalled an incident at the #2 Blowing Engine House when a co-worker crawled inside the air cylinder for repairs and swelled up enormously in the 180-degree heat. He had to be cooled with air hoses until he shrank sufficiently to fit back out the porthole exit so he could be treated for heat exhaustion. Steelworkers were dragged under railroad cars, fell from collapsing scaffolds or unsafe cranes, lost their hearing, broke bones, and suffered agonizing burns from "cobbles" or runaway pieces of red-hot steel in the rolling mills. In 1969 alone, the Lackawanna clinic receiving the plant's injured logged 58,956 emergency visits, considerably more than the largest hospitals serving the Buffalo area. Small wonder that workers refused to yield on past practices when they believed management would sacrifice their lives and health for production.

The sudden pressure to reduce crew sizes contrasted dramatically with the growth of salaried management and supervisory positions. Retired steelworker Joseph Kozak told the *Buffalo News* in 1983 that "At first we didn't need foremen. They told us what to do at the start of the shift and we did it. Then things began changing, and we had more chiefs than Indians." By 1977 3000 of Lackawanna's 11,500 employees were outside the USW jurisdiction. As of 1979 the corporation employed 23,000 salaried personnel but only 74,700 hourly or tonnage workers (including clerical staff). Though there were some cutbacks in salaried positions during 1977 and 1981, it was not until Chairman Trautlein began swinging his axe out of desperation in 1982 that casualties among Bethlehem's salaried and managerial ranks began approaching the mortality rates among the blue-collar workers.

At Lackawanna the dominant note of the 1970s and early 1980s was uncertainty. Over the years capacity was cut back, then marginally restored; facilities were idled, then reopened; employees were laid off, then recalled — all in response to the corporation's quarter-by-quarter income estimates. Art Sambuchi maintained that the company had been discussing the purchase of new equipment for the 32" rail mill the day before it was permanently closed. To some extent these fluctuations represented standard operating practice in the steel industry. One worker related in 1978 that he had been laid off 13 times in the past 14 years; another figured that though he had spent 24 years at the plant, his full-time work added up to only 10 years because of layoffs and other irregular employment.

A long strike at General Motors in 1970 and an inventory buildup in anticipation of a strike at the end of the 1971 steelworkers' contract masked the immediate effects of Bethlehem's announced plans to cut jobs and production at Lackawanna. Nevertheless, rumors were already circulating that some top executives at Bethlehem's home office advocated closing Lackawanna outright, and by the end of 1970 there were sizeable numbers of USW leaders ". . . who believe that Bethlehem plans for the middle of 1971 are not linked to contract negotiations but, instead, reflect a realignment of production plans within the entire Bethlehem system as a result of the company's construction of its new mill at Burns Harbor."

View of ruins of B furnace looking South along Blast Furnace Row (1985).

Management, whose investment priorities generated much of Lackawanna's plight, persisted in stigmatizing the plant by judging it against the performance of more favored facilities. In October, 1971 General Manager Robert Jordan announced that Lackawanna's slow recovery from a shutdown related to national contract negotiations was ". . . due to inability to compete with sister plants." Bethlehem Chairman Stewart S. Cort remarked in 1972 that in light of cooperation from union officials and local politicians, Bethlehem would keep Lackawanna's "employment level as high as can be supported by the business the plant is able to attract in competition with other producers." A company spokesman commenting on prospects for modernization of the New York facility in 1979 felt obliged to point out that "There is only so much capital, and Lackawanna will be competing for it with Burns Harbor, Sparrows Point, and other plants."

Even after the second wave of cutbacks and layoffs in 1977, though, the employment trends at Lackwanna did not immediately enter a steady course of decline. By April, 1978 3170 employees had received pink slips. From March through September of 1979 employment stabilized at 9500 or 1500 more jobs than the level projected in 1977. Plant employment totaled 7700 in January, 1980, 5900 in July, 1980, 7500 (including 2000 salaried) in August, 1981, and 6300 (with 2000 on layoff) as of November, 1981. If the handwriting was indeed on the wall for the Lackawanna plant, it was difficult to decipher the contradictory messages being emitted by the home office. Local management and workers, adversaries on most issues, shared the prevailing uncertainty about what the future would bring.

In the end, the decision to terminate steelmaking at Lackawanna rested with the company. The extinction of the facility was the final exercise of a corporate power that remained structurally intact throughout the history of 20th-century steelmaking but which had shifted priorities back to those of the early Lackawanna Steel era — profits before production.

A *New York Times* reporter compared the 1983 shutdown to the death of an aged parent or comatose patient — unspeakably painful but not utterly shocking.

View looking past remains of Bessemer building and #2 open hearth toward C furnace (1985).

Tom Hughes, an electrician in the 44" mill, evaluated the importance of the plant in people's lives:

> It's a way of life, and for 20 years you've been working swing shifts — night, days and afternoons — all in one week.
>
> Your family life is built around the steel plant, and it's something that after 20 years you get accustomed to. I really don't know how I would survive on a steady 9 to 5, Monday to Friday. It's something I've never had.

While the experience of working in steel had different meanings for every individual, their jobs were all stamped with a common imprint derived from the intense character of the production process itself. The history of the vanished workplace at Lackawanna is the story of those who labored there and who deserve now to be remembered.

BIBLIOGRAPHY

Primary Sources

Lackawanna Steel Workers Organizing Committee, Plant Grievance Records (1940-1942). BECHS

Donald K. Landsittel Collection. BECHS

Michael Schwartz, comp. "*Fortune* 500 Corporations and Their Directors, 1969," computer-generated interlock matrix and cross-index. SUNY Stony Brook, 1970.

F.G. Slagel to H. Shumway Lee. "Hours and Conditions of Labor, Lackawanna Plant," 28 March 1912. BECHS Mss. A 64-142, Box 29.

Secondary Sources

Allen, Robert C. "The Peculiar Productivity History of American Blast Furnaces, 1840-1913," *Journal of Economic History*, XXXVII/3 (Summer, 1977), pp. 605-633.

Bell, Lady. *At the Works: A Study of a Manufacturing Town.* London: Edward Arnold, 1907.

Bell, Thomas. *Out of This Furnace.* rpt. ed. Pittsburgh: University of Pittsburgh Press, 1976.

Berle, Adolph A. and Gardiner C. Means. *The Modern Corporation and Private Property.* rev. ed. New York: Harcourt, Brace & World, Inc., 1968.

Brody, David. *Steelworkers in America: The Nonunion Era.* New York: Harper & Row, 1969.

Brooks, Robert R.R. *As Steel Goes . . . Unionism in a Basic Industry.* New Haven: Yale University Press, 1940.

Bryant & Stratton College, Industrial Research Class. *A Study of Steel: A Portion of the Reports of a Visit to the Plant of the Lackawanna Branch of the Bethlehem Steel Company.* Buffalo: Bryant & Stratton College, n.d.

Butler, Joseph G., Jr. *Fifty Years of Iron and Steel.* Cleveland: Penton Press, 1920.

Camp, J.M. and C.B. Francis. *The Making, Shaping and Treating of Steel.* Pittsburgh: Carnegie Steel Company, 1920.

Davis, Horace B. *Labor and Steel.* New York: International, 1933.

Fitch, John A. *The Steel Workers.* rpt. ed. New York, Arno Press, 1969.

____________ "The Human Side of Large Outputs: Steel and Steel Workers in Six American States. Lackawanna — Swamp, Mill, and Town," *The Survey*, XXVII (October 1911-March 1912), pp. 929-945.

Fritz, John. *Autobiography of John Fritz.* New York: American Society of Mechanical Engineers, 1912.

Green, Mark and John F. Berry. *The Challenge of Hidden Profits.* New York: William Morrow and Company, Inc., 1985.

Hessen, Robert. *Steel Titan: The Life of Charles M. Schwab.* New York: Oxford University Press, 1975.

Hogan, William T. *The Economic History of the Iron and Steel Industry*, Vol. II. Lexington, MA: D.C. Heath & Co., 1971.

Interchurch World Movement, Commission of Inquiry. *Report on the Steel Strike of 1919.* New York: Harcourt, Brace and Howe, 1920.

Jeans, J. Stephen. *American Industrial Conditions and Competition: Reports of the Commissioners Appointed by the British Iron Trade Association to Enquire into the Iron, Steel, and Allied Industries of the United States.* London: Offices of the British Iron Trade Association, 1902.

Lackawanna Steel Company. *Annual Report.* 1909, 1921.

McDonnell, James R. "Treating Men Like Dirty Dogs," *New York Folklore*, X/3-4 (Summer-Fall, 1984), pp. 65-75.

Metzgar, Jack. "Plant Shutdown and Worker Response," *Socialist Review*, Vol. 10, No. 5 (Sept.-Oct., 1980), pp. 9-49.

Mizruchi, Mark S. *The American Corporate Network, 1904-1974.* Beverly Hills, CA: Sage Publications, 1982.

Moody's Industrial Manual. New York: Moody's Investor Services, Inc., 1930-1980.

Moody's Manual of Railroad and Corporation Securities. New York: Moody Manual Company, 1900-1918; Poor's Publishing Company, 1919-1924.

Moskowitz, Milton and Robert Levering. *Everybody's Business: An Almanac.* San Francisco: Harper & Row, Publishers, 1980.

Reitell, Charles. *Machinery and its Benefits to Labor in the Crude Iron and Steel Industries.* Menasha, WI: George Banta Publishing Company, 1917.

Schroeder, Gertrude G. *The Growth of Major Steel Companies, 1900-1950.* Baltimore, MD: Johns Hopkins Press, 1953.

Sonquist, John and Thomas Koenig. "Interlocking Directorates in the Top U.S. Corporations," *Insurgent Sociologist*, Vol. V, No. III (Spring, 1975), pp. 196-229.

Spencer, Charles. *Blue Collar: An Internal Examination of the Workplace.* Chicago: Lakeside Charter Books, 1977.

Stone, Katherine. "The Origins of Job Structures in the Steel Industry," *URPE Review of Radical Political Economics*, VI/2 (Summer, 1974), pp. 61-97.

Strohmeyer, John. *Crisis in Bethlehem.* Bethesda, MD: Adler & Adler, 1986.

United States Steel Corporation. *The Making, Shaping and Treating of Steel.* Pittsburgh: United States Steel Corp., 1957.

Urofsky, Melvin I. *Big Steel and the Wilson Administration: A Study in Business-Government Relations.* Columbus, OH: Ohio State University Press, 1969.

Walker, Charles R. *Steel: The Diary of a Furnace Worker.* Boston: Atlantic Monthly Press, 1922.

Warren, Kenneth. *The American Steel Industry, 1850-1970.* Oxford: Clarendon Press, 1973.

White, Victor. *Peter Domanig in America: Steel.* Indianapolis: Bobbs-Merrill, 1954.

Who Was Who in America, 1897-1942. Chicago: A.N. Marquis & Company Publishers, 1943.

Who's Who in America. Chicago: A.N. Marquis & Company Publishers. 1903-1980.
Who's Who in New York City and State. New York: L.R. Hamersley & Company, 1929.

Periodicals & Newspapers

"The Lackawanna Steel Company: Newspaper Clippings, April, 1899-March, 1911." Buffalo and Erie County Public Library Scrapbook.

Buffalo Courier-Express, Buffalo Evening News, Buffalo Labor Journal, Buffalo Live Wire, Business Week, Iron Age, Iron and Steel Engineer, Iron Trade Review, New York Times, Steel, Wall Street Journal

Photographic and Graphic Credits

Buffalo and Erie County Historical Society: Front Cover, Title Page, 6(u,l), 16(u,l), 18(u,l), 20(u,l), 24(u), 32(u), 36(l), 38(l), 40(u,l), 44, 46(u,l), 50, 62(u,l), 66(u), 70(l), 72, 74(u), 78, 100(l), Back Cover.
Ronald J. Colleran/*Buffalo News:* 116(u), 118, 120(u).
Courier-Express Collection: 70(u), 108, 116(l), 120(l), 124.
Robert L. Smith/*Courier-Express:* 126(u,l).

Hamburg Historical Society: 66(l), 68(u,l), 74(l), 98(u,l), 100(u), 102(u,l), 104(u,l).

New York State Museum Archives: 22(u,l), 24(l), 26, 28(u,l), 30(u,l), 32(l), 36(u), 38(u).

Historic American Engineering Record:
Jet Lowe: 64(u,l), 86(u,l), 88(u,l), 90(u,l), 92(u,l), 130(u,l).
Francisco Hidalgo-Romero: 14, 58, 60
Francisco Hidalgo-Romero, Carolyn Honan, Rob Hayes: 82, 84.

Publication Designer: Eileen Baxter Toro
Printed by Partners' Press, Buffalo, NY

Front Cover: Teeming ingots (1956).
Title Page: First Bessemer blow, 20 October 1903.
Back Cover: Last turn in 44" mill, 1983.